Lecture Notes in Artificial Intellige

Edited by R. Goebel, J. Siekmann, and W. Wahlster

Subseries of Lecture Notes in Computer Science

Max Bramer (Ed.)

# Artificial Intelligence

## An International Perspective

 Springer

Series Editors

Randy Goebel, University of Alberta, Edmonton, Canada
Jörg Siekmann, University of Saarland, Saarbrücken, Germany
Wolfgang Wahlster, DFKI and University of Saarland, Saarbrücken, Germany

Volume Editor

Max Bramer
University of Portsmouth, School of Computing
Lion Terrace, Portsmouth, Hants PO1 3HE, UK
E-mail: max.bramer@port.ac.uk

Library of Congress Control Number: 2009931480

CR Subject Classification (1998): I.2.3, I.2.6, I.2, D.1.6, F.4.1, J.3, J.4

LNCS Sublibrary: SL 7 – Artificial Intelligence

ISSN  0302-9743

ISBN  978-3-642-03225-7 Springer Berlin Heidelberg New York

springer.com

© IFIP International Federation for Information Processing 2009

Typesetting: Camera-ready by author, data conversion by Scientific Publishing Services, Chennai, India
Printed on acid-free paper      SPIN: 12700295      06/3180      5 4 3 2 1 0

# Preface

This edited collection of papers entitled *Artificial Intelligence: An International Perspective* originated with a proposal by the President of the International Federation for Information Processing (IFIP), Professor Basie von Solms, for the IFIP Technical Committees to publish 'position papers' on their technical areas.

Each Technical Committee has members nominated by national computer societies worldwide as their representatives. In the case of the Technical Committee on Artificial Intelligence, which I chair, the committee has members from over 35 countries and organizes a program of events on a worldwide basis through its six working groups. Like all IFIP Technical Committees, it is well placed to give an international perspective on its specialist area.

Artificial intelligence (AI) is a rapidly growing inter-disciplinary field with a long and distinguished history that involves many countries and considerably predates the development of computers. It can be traced back at least as far as ancient Greece, where Homer's *Iliad* (ca. 850 BC) told the story of how Hephaestus, the god of fire, having been cast out of Olympus creates artificial attendants to assist in his forge. The text states: "They are golden, and in appearance like living young women. There is intelligence in their hearts, and there is speech in them and strength, and from the immortal gods they have learned how to do things." Modern technology has not yet caught up with Homer, although robots in a variety of shapes and sizes are now commonplace. Five hundred years later, the attempt by the philosopher Aristotle (384 BC – 322 BC) to classify all valid forms of argument used by humans into 19 types known as 'syllogisms' can now be seen as pioneering work on understanding human reasoning in a field that was not even to be named for more than another 2,000 years.

"El Ajedristica," the first automaton to perform a non-trivial task (playing an endgame in chess) was built as long ago as 1890 by the Spanish inventor Luis Torres y Quevedo. This followed a famous and long-running hoax where a chess-playing machine was created in 1769 by Baron Wolfgang von Kempelen, a member of the court of Empress Maria Theresa in Vienna, and was exhibited around the world for the next 80 years. Dressed in the style of a Turk and seated at a cabinet with a chess board fixed to its top surface, it played and defeated opponents that included Napoleon, Frederick the Great, Benjamin Franklin and, it is said, Charles Babbage, until the secret was discovered: there was a small human chess player concealed inside the cabinet at which the Turk sat who operated the Turk's hand by means of a pantograph.

In modern times, the celebrated computing pioneer Alan Turing was a leading advocate of AI and made significant contributions to the field in the 1940s and 1950s before his untimely death in 1954. In 1950 he proposed the imitation game (now known as the 'Turing Test') by means of which it might be said that a machine was intelligent. In the same paper he said that "I believe that at the end of the [twentieth] century … one will be able to speak of machines thinking without expecting to be contradicted." Turing's timing turned out to be inaccurate, but the overall prophesy seems to be coming ever closer.

The original concerns of the field of AI (a name coined in 1956) were predominantly tasks such as playing games, solving puzzles, understanding natural language and automating mathematical theorem proving. Now that a computer has convincingly beaten the world chess champion, interest in that area has waned but in return there is commercial interest on a scale that could never have been anticipated in the past, with many hundreds of so-called expert systems (predominantly but not wholly rule based) developed in the 1980s and 1990s and since then many other commercial applications, increasingly involving the techniques of machine learning.

Today the field is large and expanding with many of the concerns of its practitioners very far from the original core areas. The time when any single volume could realistically claim to give an overview of the entire field has long passed. Instead this collection of papers aims to present an international perspective on the field from the viewpoint of expert members of Technical Committee 12, its Working Groups and their colleagues. The authors come from ten countries across three continents.

Three of the papers give overviews of work in countries (France, Italy and Chile) whose publications are not always accessible to an English-speaking audience. The inclusion of Chile in this list may seem surprising but that country has a long involvement in AI, most famously for a project which aimed to apply AI and cybernetic theory to government, which was begun by President Salvador Allende as long ago as 1971 and terminated prematurely by the coup against his government in September 1973. The other nine papers describe important relatively new or emerging areas of work in which the authors are personally involved. They are "Text and Hypertext Categorization"; "Autonomous Systems"; "Affective Intelligence"; "AI in Electronic Healthcare Systems"; "Artifact-Mediated Society and Social Intelligence Design"; "Multilingual Knowledge Management"; "Agents, Intelligence and Tools"; "Intelligent User Profiling" and "Supply Chain Business Intelligence". They provide an interesting international prospective on where this important field is going at the end of the first decade of the twenty-first century.

March 2009                                                    Max Bramer
                    Chair, IFIP Technical Committee on Artificial Intelligence
          Professor of Information Technology, University of Portsmouth, UK
                                              www.maxbramer.org

# Table of Contents

# Artificial Intelligence and Intelligent Systems Research in Chile

John Atkinson[1] and Mauricio Solar[2]

[1] Department of Computer Sciences, Universidad de Concepción, Chile
atkinson@inf.udec.cl
[2] Department of Informatics, Universidad Técnica Federico Santa María, Chile
msolar@inf.utfsm.cl

**Abstract.** Worldwide Artificial Intelligence research has witnessed fast and growing advances. These contributions mainly came from first-world nations as other research priorities and needs have been undertaken by less-developed countries. Nevertheless some Latin American countries have put significant efforts into AI research so as to advance in the state-of-the-art at international levels. This paper describes the history, evolution and main contributions of Chile to AI research and applications.

## 1 Introduction

In 1947 Alan Turing predicted that there would be intelligent computers by the end of the century. Hence he proposed an intelligence test which allows us to assess a machine as intelligent in his classic 1950 article "*Can a machine think?*". The term Artificial Intelligence (AI) arose for the first time in a conference at Dartmouth College in 1956 on machine intelligence which gathered the most renowned scientists such as John McCarthy, Marvin Minsky, Claude Shannon, Allen Newell and Herbert Simon. One of these, Herbert Simon predicted in 1965 that by 1985, machines will be capable of doing anything a man can do. At the same time Dreyfus argued against the possibilities of AI. Furthermore, Marvin Minsky from MIT predicted in 1967 that within a generation, the problem of creating AI will be substantially solved.

In the early sixties, neural network research started to spread across the most important laboratories and universities in the world. Despite the significant efforts and funding provided by public and private institutions, the famous 1969 monograph entitled *Perceptrons* [10] showed that it was impossible for these classes of neural network to learn an XOR function. Minsky and Papert conjectured incorrectly that a similar result would hold for a Perceptron with three or more layers. Three years later Stephen Grossberg published a series of papers introducing neural networks capable of modeling differential, contrast-enhancing and XOR functions. Nevertheless the often-cited Minsky & Papert paper caused a signifi-

M. Bramer (Ed.): Artificial Intelligence, LNAI 5640, pp. 1–10, 2009.

cant decline in interest and funding of neural network research. Thus, the first twenty years of AI research worldwide was characterized by high expectations in a short time.

Developing countries were not supposed to get involved in AI research in a timely fashion as renowned researchers claimed the majority of the open problems may be solved in a too short period of time. Fortunately, it was not so as a couple of South American countries such as Brazil and Chile got involved in AI in the early 80s. In Chile, the eighties witnessed the beginnings of research and development in AI. Early applications included contributions of AI in diagnostic systems in the mining industry and significant achievements on practical applications in engineering and science. In particular, strategic areas for the country were given high priority including mining, forestry and industrial automation.

A resurgence of neural networks research worldwide in the 1980s also encouraged national researchers to pursue advanced research and applications of neuromimetic systems. Some of these milestones had a significant impact on the economic resources a developing country is willing to spend on Research and Development (R&D). Since then, Computer Science (CS) research in Chile has been reinforced aiming to hire more researchers, enroll more graduate students, improve and spread research initiatives throughout the country, and increase the number of R&D projects. In the scientific area, Chilean CS shows an increase in scientific production by papers published in high quality indexed journals, renowned conferences, and ACM & IEEE conferences, etc. Most of the funding for R&D in Chile comes from the government and a few research contracts with private industries (27% of the money spent in R&D). Nevertheless, the major research funding institutions are public, including CONICYT (The Chilean NSF equivalent), FONDEF (CONICYT's funding unit for technological research in industry), CORFO (Ministry of Economy's Production Development Corporation) and MIDEPLAN (Ministry of Planning).

In this chapter, the evolution of R & D activity on AI by Chilean scientists and developers is briefly discussed and the main impacts are highlighted.

## 2    The Beginnings of Research and Development in AI in Chile

As for the majority of the AI activity worldwide, Chilean research on AI has historically had two main working focuses: Foundations and Engineering. The first stream involved seminal work by biologists, psychologists, mathematicians and linguistics on the basis of cognition, knowledge acquisition and perception. A second stream included researchers and developers, mainly from universities and research centers, working on AI applications in industry and science.

On the AI applied side, research has mainly been carried out in universities, where several paradigms can be found including *Artificial Intelligence* (mainly in CS departments), *Computational Intelligence* (mainly in Electrical and Electronic Engineering departments) and others (linguistics, psychology). A good exception outside the academic world is a project which aimed to apply AI and Cybernetic theory to the government. In 1971, an innovative system of cybernetic information management and transfer began developing in Chile during the government of President Salvador Allende; the CYBERSYN project, cybernetic synergy, information and control system. In Chilean State owned companies a system for capturing, processing and presenting economic information to be managed in "quasi" real time, based on a convergence of science, technology, politics and cybernetics, became an absolute pioneer in the application of a cybernetic model in mass socio-economic contexts. The economic system of the Allende Government, after annexing and nationalizing diverse State companies, was faced with the necessity to coordinate information regarding state companies and those that had been recently nationalized, so it required the creation of a dynamic and flexible system for proper management of the companies. In 1970 Fernando Flores was appointed General Technical Director of CORFO, and was responsible for the management and coordination between nationalized companies and the State. He had known the theories and solutions proposed by British scientist Stafford Beer since he was an engineering student, and subsequently in the course of his professional relationship with SIGMA, the Beer consultancy firm. He wrote to Stafford Beer inviting him to implement in Chile VSM (the Viable System Model), which had been developed in Beer's "The Brain Of The Firm". Beer accepted immediately, and the project entered its development stage in 1971.

In the early eighties, the Chilean scientist Dr. Fernando Flores working with Terry Winograd from Stanford University, proposed a new approach to understanding what computers do and how their functioning is related to human language, thought, and action. His influential work entitled *"Understanding Computers and Cognition"* [20] was a worldwide contribution to understand social networks and commitment nets in companies. This is based on the *Speech Acts* theory and has impact on designing intelligent systems, effective human communications, etc. It is a broad-ranging discussion exploring the background of understanding in which the discourse about computers and technology takes place. It represents an important contribution to the research about what it means to be a machine, and what it means to be human. Software systems using his research were marketed as The *Coordinator* (http://www.actiontech.com/).

Although practical applications of AI in Chile started in the early 80s, some underlying research on origins of intelligence (and its biological basis) was long before then. Research on origins of intelligence and language use (*languaging*) is due to the world renowned Chilean Scientist *Humberto Maturana*. Biologist Maturana is best known for his (re)definition of Autopoiesis [9] which has a significant impact on the understanding of natural and artificial autonomous systems. Autopoiesis theory featured two referents: a set of interconnected ideas aimed to provide definitions and explanations for life and biological phenomena. (aka. "Autopoietic theory"), and a central concept in this set of ideas; the defining property of living systems.

With roots in Cybernetics, Maturana's work influenced various fields; hot topic in *Artificial Life and embodied* dynamical approaches to cognitive science. Its basic axioms involved two key principles:

- *Structural determinism*. The dynamics of a system are determined only by its own structural composition, following operational laws (the laws of physics).

- *Everything said is said by an observer*. It is not possible to do science without a point of view and a language that influences what the observer chooses to distinguish in her/his observations.

Maturana's theory applies science, especially what is known of neural systems, to philosophical questions about human perception and understanding (Autopoiesis and Cognition). This addresses the origin of life and continues through the development of language in humans [9]. Furthermore, A Maturana's colleague, Francisco Varela provided the biological basis for designing complex systems, artificial life and embodied intelligence. For Varela, the core of intelligence and cognitive abilities is the same as the capacity of living which gives rise to modern *Artificial Life* [18, 19] with applications to control theory, robotics and the shift towards biologically inspired notions of viability and adaptation, situatedness, etc. He addressed *bottom-up theories of artificial intelligence* and explored what can be learned from simple models such as insects about the cognitive processes and characteristic autonomy of living organisms. Dr. Varela first introduced the notion of "*Emergence*" (connectionism): many cognitive tasks (such as vision and memory) seem to be handled best by systems made up of many simple components, which, when connected by the appropriate rules, give rise to global behaviours.

# 3     AI at Universities

An interesting feature of AI research in Chile was that this aimed to solve practical problems in industry. For example, early development of expert systems in Chile was applied to the mining industry (one of the most important in the World), process control, and medicine. The first expert system (ES) outside the mining area was capable of diagnosing faults in motor pumps for paper companies, other applications included an expert controller for Semi-Autogenous Grinding [11]; a configuration system for climatic testings [16], and Fuzzy ES for automation. Later, a group of Chilean researchers pioneered the foundations for essential hypertension treatment using AI techniques [8].

In the middle 80s, several Chilean experts pursuing graduate studies returned to the country. They were mainly specialized in ES, neural network systems, evolutionary computation and heuristic optimization. This drove the development and application of AI technology in the Chilean industry. The first Chilean graduates were from Georgia Tech (USA), King's College London (UK), UFRJ in Brazil,

University of Toronto (Canada), etc. In the last five years, graduates returning to the country came from INRIA (France), University of Edinburgh (UK), University of Cambridge, Carnegie Mellon University (USA), etc.

Currently, CS departments of all major universities in the country have a research area in AI. There is no common thematic or central AI organization in the country, but most of the researchers are members of the Chilean Computer Science Society. An overview of the main activities in some of the major institutions are highlighted including those by Pontificia Universidad Catolica de Chile (PUC), Universidad de Concepción (UCO), Universidad de Santiago de Chile (USACH), Universidad de Chile (UChile) and Universidad Técnica Federico Santa Maria (UTFSM).

## 3.1    AI at PUC

The main research activities in AI at PUC focus on machine learning, computer vision and autonomous robotics. In the early 80s, a significant development of logics and theorem solving was carried out by [5, 6, 7]. Furthermore, PUC pioneered the use and application of ES technology in industry. This led to the first Expert System company in the early 80s: *SOLEX*, and the development of applications in industrial planning, optimization, and heuristics. By the early 90s, milestones included influential work on logic for knowledge representation by Professor Leopoldo Bertossi (now at Carleton University, Canada), and *Situation Calculus* by Professor Javier Pinto (RIP) [14, 15]. Some other recent developments included Robotics and Probabilistic Reasoning, Agent planning under uncertainty using logic-based programming, Dynamic Surveillance using Unmanned Aerial Vehicles and Planning visual navigation for mobile robots. Applied machine learning has also been a recent focus for detecting rare objects in huge astronomical databases using data mining technologies.

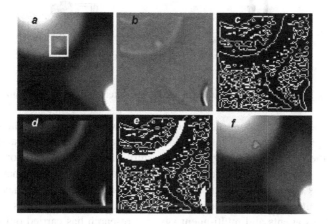

**Fig. 1.** Flaw Detection in uncalibrated sequence of images

An example of computer vision research is the automated inspection using un-calibrated image sequences. This technology integrates state-of-the-art image processing and pattern recognition techniques (see Fig. 1).

## 3.2   AI at UCO

Early AI research was marked by applied work on ES in medium-size companies and knowledge representation theories for adaptive ES. Recent work has contrib-uted significantly to advancing applied AI on areas such as Natural Language Processing (NLP), Multi-agent systems, evolutionary computation, pattern recog-nition and intelligent optimization. The work at UCO is the Chilean leader on NLP and language technology including text mining, natural-language dialogue systems and semantic processing. Some of this work has been published in the most prestigious international journals [1, 2, 3, 4]. NLP work has been pursued jointly with researchers in the Department of Linguistics for the last 10 years. A current worldwide contribution [3] involves a new model to filter information from the Web using natural-language dialogue systems (see Fig. 2).

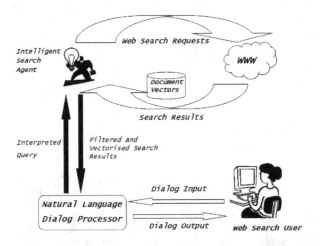

**Fig. 2.** A New Natural-Language Dialogue Model for Filtering information from the Web.

In the new paradigm, users do not get overloaded with information when search-ing on the web. Instead, they interact with a dialogue system which attempts to understand their requests. Dialogue goes on until most of the user's communica-tion goals have been met. The model is then capable of filtering information from the web that is of specific interest to the user. Multiple experiments have shown that this new filtering model outperforms state-of-the-art search systems.

Autonomous agents and multi-agent systems research has carried out work on intelligent search on the web, negotiation strategies for dynamic formation of agents, and multi-agent simulation. Furthermore, the Department of Computer

Sciences has one of the two Chilean robotic soccer teams participating in the world robotic soccer competition *RoboCup*. A recent achievement is in intelligent robotic soccer in which team formation strategies are dynamically generated by using neural networks learning from the game conditions and the policy provided by a human coach.

The University has also contributed influential work on AI applications in foreign language learning and intelligent tutoring systems. Recent work focuses on automatic feedback generation using NLP technology for intelligent tutoring of language learning. Significant research has also been pursued on pattern recognition and its applications in Bioinformatics. In particular, new clustering and recognition methods for microarray image analysis and DNA processing.

## 3.3    AI at USACH

Current research has mainly focused on AI applications in medicine and practical optimization in industry. Research groups at USACH along with other universities pioneered a national effort to advance research and applications of neural network based methods in industry in the middle eighties. Significant applied AI work has also been in autonomous robotics applications for the Chilean mining industry. Recently, an important effort has concentrated on using machine learning techniques for Business Intelligence applications.

In Heuristics and Optimization, the focus is on solving hard problems in engineering and industry [12, 13]; Metaheuristics (i.e., Genetic Algorithms) for NP-hard problems; approaches to optimization using parallel and distributed computing technology [17]. Optimization using other techniques such as neural networks includes influential applied work on neural network models for pattern recognition, time-series prediction, metal image recognition and Intelligent Signal Processing. The Department of Computer Science Engineering has a key national role in biomedical applications in different areas such as data mining in the health area, biological systems analysis, neural nets and Bayesian methods for assessing health technologies.

## 3.4    AI at UChile

The Computer Science Department at University of Chile has no significant research contributions in AI. However, a small research group spread over several departments (electrical engineering, computer science, and industrial engineering) has been engaged on AI related research. For example, some significant work has been developed in the areas of dynamic data mining and feature selection using machine learning methods. Furthermore, important research on formal Logic and the Semantic Web including a metadata model for describing and querying on the semantic web have been pursued by researchers in logic, mathematics and computer science. Recently, the Department of Electrical Engineering has led the application of AI techniques on intelligent robotic systems.

### 3.5   AI at UTFSM

In addition to the academic AI research in universities some nationwide milestones have been witnessed across universities and companies. For example, Chile
was the first South American country to introduce ES technology for fault diagnosis and knowledge acquisition into the Mining industry (80s). The first experimental and practical Natural Language Interfaces for business purposes were developed in early 80s. The first International Symposium on Artificial Intelligence in
South America was held in Valparaiso, Chile (early 80s) in cooperation with the
French Government. In early 90s the first practical mobile robotics applications
for mine operations were deployed. In the middle 90s, a consortium of Universities brought Neural Nets Engineering into the productive area funded by
FONDEF. Results included applications in pattern classification; Automatic Visual inspection; Robotic Manipulation; and Financial analysis. Recently, the government granted a million US dollar project to develop the first national Univer-
sity-Industry consortium for designing Unmanned Aerial Vehicles for surveillance
and monitoring applications (2005).

AI work at UTFSM has been conducted in the Department of Computer Science and Department of Electronic Engineering. Main research areas have focused
on pattern recognition, evolutionary computation and heuristic optimization and
neural networks. New ensemble machine learning techniques have been developed
to deal with data-intensive tasks such as image processing, climate analysis, etc.
Statistically-motivated methods have also been designed to improve the parameter
setting task in evolutionary computation methods. Recently, important ongoing
work has been developed for intelligent robotics in order to provide evolutionary
mechanisms of navigation and object tracking.

## 4    Conclusions

Chilean AI basic and applied research has shown important contributions in industry, science and technology both at national and international levels. Keeping in
mind the population of Chile (a 16 million people country), its activities are even
comparable in term of *per capita* activities to other bigger developing countries
such as Brazil.

However, massive introduction of AI technology into private companies is still
in its early stages. Data mining technologies, fraud detection, industrial-strength
optimization, etc are some exceptions. As biotechnology, and in particular, bioinformatics becomes more popular, AI techniques for processing massive genomic
data, image sequence and DNA information are a must. There are huge gaps in
education as being recognized by educational authorities. Designing and applying
AI systems for improving the learning/teaching process can address the main issues.

# References

[1]   Atkinson, J., Mellish, C.: A Semantically-guided and Domain-independent Evolution-ary Model for Knowledge Discovery from Texts. IEEE Transactions on Evolutionary Computation, pp. 546–560 (December 2003).

[2]   Atkinson, J., Mellish, C.: Combining Information Extraction with Genetic Algorithms for Text Mining. IEEE Intelligent Systems and Their Applications, 2–30 (2004)

[3]   Ferreira, A., Atkinson, J.: Intelligent Search Agents using Web-driven Natural Lan-guage Explanatory Dialogs. IEEE Computer, Special Theme on Intelligent Search, 44–52 (Oct. 2005)

[4]   Ferreira, A., Moore, J., Mellish, C.: A Study of Feedback Strategies in Foreign Lan-guage Classrooms and Tutorials with Implications for Intelligent Computer-Assisted Language Learning Systems. Journal of Artificial Intelligence in Education 17, 389–422 (2007)

[5]   Fuller, D., Abramsky, S.: Partial Evaluation and Mixed Computation of PROLOG Programs. In: Proceedings of the Workshop on Partial Evaluation and Mixed Compu-tation, Gl., Avernaes, Denmark (1987)

[6]   Fuller, D., Bocic, S.: Extending Partial Evaluation in Logic Programming. In: Com-puter Science: Research and Applications, Proceedings of the XI International Con-ference of the Chilean Computer Science Society, Santiago, Chile, pp. 95–107. Ple-num Publishing Corp., New York (1992)

[7]   Fuller, D., Bocic, S., Bertossi, L.: Towards Efficient Partial Evaluation in Logic Programming. New Generation Computing Journal (Feb. 1996)

[8]   Holzmann, C., Estévez, P., Acuña, G., Schifferli, G., Rosselot, E., Pérez-Olea, E.: Fuzzy Expert System Foundations for the Essential Hypertension Treatment. In: Pro-ceedings of the V Mediterranean Conference on Medical & Biological Engineering, MEDICOM-89, University of Patras, Greece, August 1989, pp. 324–325 (1989)

[9]   Maturana, H., Varela, F.: The tree of knowledge: the biological roots of humans understanding. Shambhala Publications, Boston (1987)

[10]  Minsky, M., Papert, S.: Perceptrons: An Introduction to Computational Geometry. MIT Press, Cambridge (1969)

[11]  Muñoz, A.: Expert Control for Semi-Autogenous Grinding. M.Eng thesis, Catholic University of Chile, Santiago, Chile (1988)

[12]  Parada, V., Sepulveda, M., Solar, M., Gomes, A.: Solution for the constrained guillo-tine Cutting Problem by Simulated Annealing. Journal on Computers and Operations Research 25(1), 37–47 (1998)

[13]  Parada, V., Pradenas, L., Solar, M., Palma, R.: A Hybrid Algorithm for the Non-Guillotine Cutting Problem. Annals of Operations Research 117, 151–163 (2002)

[14]  Pinto, J.: Using histories to model observations in theories of action. In: Antoniou, G., Truszczyński, M. (eds.) PRICAI-WS 1996. LNCS, vol. 1359, pp. 221–233. Springer, Heidelberg (1998)

[15]  Pinto, J.: Integrating discrete and continuous change in a logic framework. Computa-tional Intelligence 14, 39–88 (1998)

[16]  Solar, M.: An Expert System for Configuring Climatic Testings. M.Sc thesis, UFRJ, Brazil (1989)

[17]  Solar, M., Urrutia, R., Parada, V.: Parallel Genetic Algorithm to Solve the Set-Covering Problem. Journal on Computers and Operations Research 29(9), 1221–1235 (2002)

[18] Varela, F., Bourgine, P.: Toward a practice of autonomous systems: Proceedings of the first European Conference on Artificial Life (1992)

[19] Varela, F., Rosch, T.: The embodied mind: cognitive science and human experience. MIT Press, Cambridge (1992)

[20] Winograd, T., Flores, F.: Understanding Computers and Cognition: A New Foundation for Design. Greenwood Publishing Group Inc., Westport (1986)

# Text and Hypertext Categorization

Houda Benbrahim[1] and Max Bramer[2]

[1] Ernst and Young LLP, 1 More London Place, London SE1 2AF, United Kingdom
hbenbrahim@uk.ey.com
[2] School of Computing, University of Portsmouth, Portsmouth, Hants PO1 3HE,
United Kingdom
max.bramer@port.ac.uk

**Abstract.** Automatic categorization of text documents has become an important area of research in the last two decades, with features that make it significantly more difficult than the traditional classification tasks studied in machine learning. A more recent development is the need to classify hypertext documents, most notably web pages. These have features that add further complexity to the categorization task but also offer the possibility of using information that is not available in standard text classification, such as metadata and the content of the web pages that point to and are pointed at by a web page of interest. This chapter surveys the state of the art in text categorization and hypertext categorization, focussing particularly on issues of representation that differentiate them from 'conventional' classification tasks and from each other.

## 1 Introduction

Over the past two decades, automatic content-based document management tasks have received a great deal of attention, largely due to the increased availability of documents in digital form and the consequential need on the part of users to access them in flexible ways. Text categorization or text classification, which is the assignment of natural language texts into predefined thematic categories, is one such task.

Hypertext categorization is a variant of text categorization, where the documents concerned are web pages. The nature of web pages presents additional challenges when representing the information they contain, but also offers the opportunity to include information such as metadata that is unavailable with standard documents.

This chapter begins by summarizing the state of the art in text categorization and then goes on to look at hypertext categorization, concentrating particularly on the differences from standard text that derive from the special nature of web pages. The emphasis is on illustrating how the information in web pages can best be represented including information that is not available with standard text, such

M. Bramer (Ed.): Artificial Intelligence, LNAI 5640, pp. 11–38, 2009.

as metadata and information about the web pages that point to or are pointed to by a page under consideration. Once a collection of text or hypertext documents have been converted to a dataset in standard attribute/value form they can be processed by standard classification algorithms such as decision trees (Quinlan, 1986), support vector machines (Vapnik, 1995) etc. Information about these is readily available and they will not be described here.

## 2    Text Categorization

Text categorization (TC) dates back to the early 1960s, but it was not until the early 1990s that it gained popularity in the information systems area, due to the exponential growth of online text and also advances in computer hardware. TC is used in many application contexts, ranging from automatic document indexing based on a controlled vocabulary (Borko and Bernick 1963; Gray and Harley 1971; Field 1975), to document filtering (Amati and Crestani 1999; Iyer, Lewis et al. 2000; Kim, Hahn et al. 2000), word sense disambiguation (Gale, Church et al. 1992; Escudero, Marquez et al. 2000), population of hierarchical catalogues of Web resources (Chakrabarti, Dom et al. 1998; Attardi, Gulli et al. 1999; Oh, Myaeng et al. 2000), and in general any application requiring document organization or selective and adaptive document dispatching.

Until the late 1980s the most popular approach to TC was the knowledge engineering (KE) one, where a set of rules appropriate to the problem under consideration was manually created by encoding expert knowledge. Normally this approach requires a great human effort in defining the right rules. It is not flexible and a small change in the domain implies a large effort to modify the whole system. A system designed for a given domain is not adaptable for another domain and it must be redesigned from scratch with high costs in time and human effort.

From the early 1990s this approach has increasingly lost popularity, especially in the research community, in favour of the machine learning (ML) approach. In this framework, a general inductive process automatically builds an automatic text classifier by learning, from a set of previously classified documents, the characteristics of the categories of interest. The advantages of this approach are (i) classification accuracy comparable to that achieved by human experts, and (ii) a considerable saving in terms of expert manpower compared with the KE approach, since no intervention from either knowledge engineers or domain experts is needed for the construction of the classifier or for its porting to a different category set.

Developing a generic text classifier system based on machine learning (ML) techniques requires important decisions to be made about *text representation*: the way in which a document is represented so that it can be analyzed. Text representation comprises several phases:

- indexing (defines and extracts index terms that will be considered as features representing a given document)

- feature reduction (removes non-informative features from documents to improve categorization accuracy and reduce computational complexity)

- feature vector generation (represents each document as a weighted *feature vector* that can later be used to generate a text classifier or as input to the resulting classifier).

Each of these is discussed in Section 3 below.

Once the set of training documents is represented by a number of feature vectors, a machine learning algorithm can be applied to analyze it. At this stage, a *training set* of documents whose correct classifications are known is needed. The output of the learning phase is a model (called a *classifier*) that can be used later to classify new unseen documents.

The accuracy of the classifier is estimated by using it to classify a set of pre-labelled documents (called a *test set*) which are not used in the learning phase. The classifications produced are compared with those previously assigned to them (for example by human classifiers) which are treated as a gold standard.

## 2.1 Machine Learning Approaches to Text Categorization

Text classifiers developed using ML algorithms have been found to be cheaper and faster to build than ones developed by knowledge engineering, as well as more accurate in some applications (Sj and Waltz 1992). Nevertheless, ML algorithms applied to TC are challenged by many properties of text documents: a high dimensionality feature set, intrinsic linguistic properties (such as synonymy and ambiguity) and classification of documents into categories with few or no training examples.

A wide variety of learning approaches have been applied to TC, to name a few, Bayesian classification (Lewis and Ringuette 1994; Domingo and Pazzani 1996; Larkey and Croft 1996; Koller and Sahami 1997; Lewis 1998), decision trees (Weiss, Apte et al. ; Fuhr and Buckley 1991; Cohen and Hirsh 1998; Li and Jain 1998), decision rule classifiers such as CHARADE (Moulinier and Ganascia 1996), or DL-ESC (Li and Yamanishi 1999), or RIPPER (Cohen and Hirsh 1998), or SCAR (Moulinier, Raskinis et al. 1996), or SCAP-1 (Apté, Damerau et al. 1994), multi-linear regression models (Yang and Chute 1994; Yang and Liu 1999), Rocchio method (Hull 1994; Ittner, Lewis et al. 1995; Sable and Hatzivassiloglou 2000), Neural Networks (Schütze, Hull et al. 1995; Wiener, Pedersen et al. 1995; Dagan, Karov et al. 1997; Ng, Goh et al. 1997; Lam and Lee 1999; Ruiz and Srinivasan 1999), example based classifiers (Creecy 1991; Masand, Linoff et al. 1992; Larkey 1999), support vector machines (Joachims 1998), Bayesian inference networks (Tzeras and Hartmann 1993; Wai and Fan 1997; Dumais, Platt et al. 1998), genetic algorithms (Masand 1994; Clack, Farringdon et al. 1997), and maximum entropy modelling (Manning and Schütze 1999).

## 2.2    Benchmarks

To establish which classification model is the most accurate, several TC bench-marks have been developed. The best-known one is probably the Reuters corpus[1]. It consists of a set of newswire stories classified under categories related to eco-nomics. The Reuters collection has been used for most of the experimental work in TC so far.

Other collections have also frequently been used such as the OHSUMED col-lection (Hersh, Buckley et al. 1994) used in (Joachims 1996; Baker and McCallum 1998; Lam and Ho 1998; Ruiz and Srinivasan 1999), the 20-newsgroups (Lang 1995) and in (Baker and McCallum 1998; Joachims 1998; McCallum and Nigam 1998; Nigam and Ghani 2000; Schapire and Singer 2000).

## 2.3    Text Categorization Compared with Traditional Classification Tasks

Text categorization uses several of the successful classification algorithms that have been developed for other 'traditional' classification areas. However, those algorithms have to overcome some challenges caused by the key characteristics of the TC area:

- High dimensional feature space. In text categorization, the input to the learning algorithm consists of a set of all the words occurring in all the training docu-ments. A few thousand training documents can lead to tens of thousands of fea-tures. The dimensionality can be reduced using methods such as those de-scribed in Section 3.2, yet the resulting feature set will generally still be very large.

- Sparse document vectors. Even if there is a large set of features, each document contains only a small number of distinct words. This implies that document vectors are very sparse, i.e. only a few words occur with non-zero frequency.

- Heterogeneous use of terms. To cut down the size of the feature space, feature selection methods can be used to discard all the irrelevant features. However, in text classification this can lead to loss of information. Documents from the same category can consist of different words, since natural language allows the expression of related content with different formulation. There is generally not a small set of words that sufficiently describes all documents with respect to the classification task.

- High level of redundancy. In general, there are many different features relevant to the classification task and often several of those features occur in one docu-ment. This means that document vectors are redundant with respect to the clas-sification task.

---

[1]  http://www.daviddlewis.com/resources/testcollections/reuters21578/

- Noise. Most natural language documents contain language mistakes. In machine learning, this can be interpreted as noise.

- Complex learning task. In text classification, the predefined classes are generally based on the semantic understanding of natural language by humans. Therefore, the learning algorithm needs to approximate such complex concepts.

- In 'conventional' classification problems studied by researchers in Machine Learning the classifications are usually mutually exclusive and exhaustive (very good, good, neutral, bad, very bad etc.). By contrast a text document can often belong to several categories, for example 'military history', 'crime' and 'second world war'. This is generally handled by converting a categorization problem with N categories into N separate binary yes/no categorization problems ('Is this document about military history yes/no?', 'Is it about crime yes/no?' etc.). Building N classifiers not just one obviously adds greatly to the processing time required.

## 3     Text Representation

### 3.1     Indexing

Text documents cannot be directly interpreted by a classifier. An indexing procedure needs to be applied to the dataset that maps each text document onto a compact representation of its content.

The choice of a representation for text depends on what one considers to be the meaningful textual units (the problem of lexical semantics) and the meaningful natural language rules for the combination of these units (the problem of compositional semantics).

A fundamental challenge in natural language processing and understanding is that information or meaning conveyed by language always depends on context. The same word might have different meanings. Also, a phrase may have different meanings depending on the context in which it is used. Consequently, trying to represent natural language documents by means of a set of index terms is a challenging task. Different linguistic approaches try to capture, or ignore, to varying degrees meaning with respect to context. These approaches can be divided into five levels:

- *Graphemic:* analysis on a sub-word level, commonly concerning letters.
- *Lexical:* analysis concerning individual words.
- *Syntactic:* analysis concerning the structure of sentences.
- *Semantic:* analysis related to the meaning of words and phrases.
- *Pragmatic:* analysis related to meaning regarding language-dependent and language-independent, e.g. application-specific, context.

The graphemic and lexical levels capture basically the frequencies of letter combinations or words in the documents. Text representation based on those approaches cannot deal with the meaning of documents, as there is a weak relationship between term occurrences and document content. On the other hand, the syntactic, semantic and pragmatic levels exploit more contextual information, such as structure of sentences, and lead to more complex text representation.

Choosing a suitable level of text analysis on which to base the definition of terms is a trade-off between semantic expressivity and representational complexity. A complex text representation leads to an increase in the dimension of the feature space, and with a limited number of training documents, inducing an accurate classifier is much harder. For this reason, in practice simple term definitions are dominant in text representation. The *n-grams* (overlapping and contiguous n letter subsequences of words) approach has often been used for indexing (De Heer 1982; Cavnar and Trenkle 1994; Tauritz, Kok et al. 2000). The advantage of n-grams is that the set of possible terms is fixed and known in advance (using only the 26 letters of the English alphabet and n=3, there are $26^3=17,576$ distinct trigrams). Furthermore, n-grams are language independent and are quite robust to both morphological and spelling variations and mistakes. N-grams are easy to calculate but the resulting representation is difficult to analyze by humans.

The most widely-used approach for indexing is the use of words, known as *tokens*. In this approach, the sequence in which the words appear in a document and any structure of the text are ignored. Effectively the document is treated as a *bag-of-words* (Lewis and Ringuette 1994). This term definition is language independent and computationally very efficient. However, a disadvantage is that each inflection of a word is a possible feature and thus the number of features can be unnecessarily very large. With a bag-of-words representation, we cannot tell (for instance) if the phrase "machine learning" exists within the document, or if the two words appeared unrelated to each other (Cohen and Singer 1998). Despite ignoring this information, classifiers using this representation perform well enough in comparison with ones that take word order into account.

Another approach is to use phrases, i.e. combining many words as one index, for example "artificial intelligence" or "data mining" (Fuhr and Buckley 1991; Tzeras and Hartmann 1993; Schütze, Hull et al. 1995). These indexes can be generated either manually or automatically. Because this process is highly domain dependent and considering all possible combinations of tokens is impossible, many algorithms exist to define phrasal indexes. Although some researchers have reported an improvement in classification accuracy when using such indexes (depending on the quality of the generated phrases), a number of experimental results (Lewis 1992; Apté, Damerau et al. 1994; Dumais, Platt et al. 1998) have not been uniformly encouraging, irrespective of whether the notion of "phrase" is motivated (i) *syntactically*, i.e. the phrase is such according to the grammar of the language (Lewis 1992); or (ii) *statistically*, i.e. the phrase is not grammatically such, but is composed of a set/sequence of words whose patterns of contiguous occurrence in the collection are statistically significant (Caropreso, Matwin et al. 2001).

(Lewis 1992) argues that the likely reason for those discouraging results is that, although indexing languages based on phrases have superior semantic qualities, they have inferior statistical qualities with respect to word-only indexing languages: a phrase-only indexing language has more terms, more synonymous or nearly synonymous terms, lower consistency of assignment (since synonymous terms are not assigned to the same documents), and lower document frequency for terms.

After tokens are extracted from files, an indexing phase follows. This consists of building a sequence of indexes based on those tokens. The way we use this set of indexes depends on whether the dictionary is already built or not, and therefore, whether the corresponding document is used for learning or for classification. When the dictionary is not yet constructed, the output of the tokenization step is merged to create a set of distinct indexes.

There are two ways in which the indexes can be chosen in order to build the dictionary. They can be selected to support classification under each category in turn, i.e. only those indexes that appear in documents in the specified category are used (the *local dictionary* approach). This means that the set of documents has a different feature representation (set of features) for each category. Alternatively, the indexes can be chosen to support classification under all categories, i.e. all indexes that appear in any of the documents in the training set are used (the *global dictionary* approach).

If the dictionary has already been created, then for each document under consideration the indexes are extracted and the ones that are not in the dictionary are omitted. This set of indexes is then used for document representation.

## 3.2    Feature Reduction

The main goal of the text classification task is to classify documents based on their contents. To do this, much of the auxiliary and collateral information inserted by writers to enrich the text exposition is not useful and on the contrary can make the decision more difficult. The system should analyze only the most informative words that can help to infer the topic of the document. Because of this, many auxiliary verbs, adverbs, conjunctions etc. are absolutely useless since they do not give a description of the topic. Also, they are often uniformly distributed over all topics and their informative contribution is uniformly spread over the classes.

A central problem in text classification using the machine learning paradigm is the high dimensionality of the feature space. There is one dimension for each unique index term found in the collection of documents and it is possible to have tens of thousands of different words occurring in a fairly small set of documents. Using all these words is time consuming and represents a serious obstacle for a learning algorithm. Standard classification techniques cannot deal with such a large feature set: not only does processing become extremely costly in computational terms, but the results become unreliable because of the lack of sufficient training data. However, many of the features are not really important for the learning task and their usage can degrade the system's performance. There is an important need to be able to reduce the original feature set. There are two commonly used techniques of feature reduction: feature extraction and feature selection.

### 3.2.1  Feature Extraction

A frequent problem with using plain words as features is that the 'morphological variants' of a word will each be considered as a separate feature, for example computer, computers, computing and compute will be considered to be four different features, and this will frequently result in the number of features being unnecessarily large.

*Feature Extraction* is the process of extracting a set of new features from some of the original features (generally to replace them) through some functional mapping. Its drawback is that the generated new features may not have a clear physical meaning.

*Stemming* is among the widely used methods for feature extraction. It aims at conflating morphological variants of words by removing suffixes and mapping words to their base form. Mapping morphologically similar words into their stem can cut down the number of features substantially, for example 'compute', 'computing', 'computer' and 'computers' might all be mapped to 'comput'.

Many strategies for suffix stripping have been reported in the literature (Lovins 1968; Petrarca and Lay 1969; Andrews 1971; Dattola 1979; Porter 1980). The most widely used stemming tools for the English language are the Porter stemmer (Porter 1980) and the word-net morphological processor (Miller, Princeton et al. 1998). Other methods include term clustering techniques (Baker and McCallum 1998; Li and Jain 1998; Slonim and Tishby 2001) and latent semantic indexing (Deerwester, Dumais et al. 1990; Schütze, Hull et al. 1995; Wiener, Pedersen et al. 1995).

### 3.2.2  Feature Selection

*Feature Selection* is the process of choosing a subset from the original feature set according to some criteria. The selected feature retains original physical meaning and provides a better understanding for the data and learning process. However, in removing terms, the risk is of removing potentially useful information on the meaning of documents, so the reduction process needs to be performed with care.

*Stop Words*

One of the most frequently used methods of reducing the number of different features in the dictionary is to remove very common words known as *stop words*. These are high frequency words that are likely to exist within all or nearly all documents of the collection regardless of their classifications. In the case of phrasal indexes, stop words should be eliminated before dictionary building.

Articles, prepositions and common verbs (i.e. the, from, do etc.) that provide structure in the language rather than content, are usually considered to be stop words.

| a | at | during | if | last | near | that |
|---|---|---|---|---|---|---|
| about | be | each | in | late | no | the |
| all | but | else | is | like | of | they |
| an | by | for | it | many | often | to |
| and | did | from | into | much | on | with |
| are | do | further | itself | more | once | which |
| as | down | get | just | must | or | whether |

**Fig. 1.** Some Frequently Used Stop Words

A list of 512-predefined English stop words can be found in (Lewis 1992). There is also the possibility of constructing a domain-dependent-list.

### Using a Thesaurus

Another important objective in the context of feature reduction is to conflate synonyms, i.e. words of the same or similar meaning. A *thesaurus* is a collection of words that are grouped by likenesses in their meaning rather than in alphabetical order. Mapping all the index terms onto the equivalent class to which they belong can thus reduce the set of features.

Another common way of using a thesaurus is to expand rather than conflate the set of features, by adding semantically related words to the set. A thesaurus is domain- and language dependent and manually constructed thesauri are thus often used. However, there are also attempts to automatically construct thesauri based on document collections.

### Document Frequency Thresholding

Another effective feature selection method is *document frequency thresholding*. Document frequency is defined as the number of documents in which a term occurs. The document frequency is computed for each unique term in the training dataset and those whose frequency is less than a predefined threshold are removed from the feature set. The basic assumption is that rare terms are either non-informative for category prediction or not influential in global performance. In either case, removal of rare terms reduces dimensionality of the feature space. Improvement in categorization accuracy is also possible if rare terms happen to be noise, e.g. spelling mistakes.

(Luhn 1958) suggested that the most relevant words belong to the intermediate interval of frequency and the irrelevant ones are out from this range.

*Inverse Document Frequency and TFIDF*

A significant problem with the Document Frequency Thresholding approach is that although it is true that rarely occurring terms are unlikely to be relevant for characterizing a particular class, terms that occur in a large portion of the document collection may not be discriminative either. The *term frequency/inverse document frequency (TFIDF)* method assumes that (common stop words having been removed) the importance of a term increases with its use frequency but is inversely proportional to the number of documents in which it appears. For a term *t*, the former is denoted by *tf(t)*, the frequency of term *t* in all documents. A possible measure for the latter is the *inverse document frequency* for term t, which is defined as

$$idf(t) = log \ (n \ / \ n(t))$$

where *n* is the number of documents in the training set and *n(t)* is the number of documents where term *t* occurs.

This leads to the term frequency/inverse document frequency (tfidf) measure:

$$tfidf(t) = tf \ (t) \ * idf(t)$$

The more important terms are those assigned higher *tfidf* values. Note that the evaluation of this measure does not depend on the class labels of the training documents.

Other more sophisticated information-theoretic functions, which use a term-goodness criterion threshold to decide about eliminating a feature, have been used in the literature. Among these are the DIA association factor (Fuhr and Buckley 1991), chi-square (Yang and Pedersen 1997; Sebastiani, Sperduti et al. 2000; Caropreso, Matwin et al. 2001), NGL coefficient (Ng, Goh et al. 1997; Ruiz and Srinivasan 1999), information gain (Lewis 1992; Lewis and Ringuette 1994; Moulinier, Raskinis et al. 1996; Yang and Pedersen 1997; Larkey 1998; Mladenic and Grobelnik 1998; Caropreso, Matwin et al. 2001), mutual information (Larkey and Croft 1996; Wai and Fan 1997; Dumais, Platt et al. 1998; Taira and Haruno 1999) odds ratio (Mladenic and Grobelnik 1998; Ruiz and Srinivasan 1999; Caropreso, Matwin et al. 2001), relevancy score (Wiener, Pedersen et al. 1995) and GSS coefficient (Galavotti, Sebastiani et al. 2000). Three of the most popular methods are descrivbed briefly below. They all make use of the class labels of the training documents.

*Information Gain (IG)*

Information gain is widely used in the field of machine learning. It measures the number of bits of information obtained for category prediction by knowing the presence or absence of a term in a document. The information gain of term t is defined to be:

$$G(t) = -\sum_{i=1}^{m} \Pr(c_i) \log \Pr(c_i) + \Pr(t) \sum_{i=1}^{m} \Pr(c_i \mid t) \log \Pr(c_i \mid t) + \Pr(\bar{t}) \sum_{i=1}^{m} \Pr(c_i \mid \bar{t}) \log \Pr(c_i \mid \bar{t})$$

where:

$Pr(c_i)$ is the probability of having class $c_i$

$Pr(t)$ is the probability of having term $t$

$Pr(c_i \mid t)$ is the probability of having class $c_i$ given that the term $t$ is observed in the document.

$Pr(c_i \mid t^-)$ is the probability of having class $c_i$ given that term $t$ is not observed in the document.

### Mutual Information (MI)

Mutual information is a criterion commonly used in statistical language modelling of word associations and related applications. If one considers the two-way contingency table of a term $t$ and a category $c$, where $A$ is the number of times $t$ and $c$ co-occur, $B$ is the number of time $t$ occurs without $c$, $C$ is number of times $c$ occurs without $t$, and $N$ is the total number of documents, then mutual information criterion between $t$ and $c$ is defined to be:

$$I(t,c) = \log \frac{Pr(t \wedge c)}{Pr(t) * Pr(c)}$$

and is estimated using:

$$I(c,t) \approx \log \frac{A * N}{(A+C)+(A+B)}$$

$I(t,c)$ has a natural value of zero if $t$ and $c$ are independent. To measure the goodness of a term in a global feature selection, we combine the category specific scores of a term in two ways:

$$I_{avg}(t) = \sum_{i=1}^{m} Pr(c_i) I(t,c_i)$$

$$I_{max}(t) = \max_{i=1}^{m} \{ I(t,c_i) \}$$

A weakness of mutual information is that the score is strongly influenced by the marginal probabilities of terms, as can be seen in this equivalent form:

$$I(c,t) = \log Pr(t \mid c) - \log Pr(t)$$

For terms with an equal conditional probability $Pr(t|c)$, rare terms will have a higher score than common terms. The scores, therefore, are not comparable across terms of widely differing frequency.

### $\chi^2$ Statistic (CHI)

The $\chi^2$ statistic measures the lack of independence between t and c and can be compared with the $\chi^2$ distribution with one degree of freedom to judge extremeness. Using the two-way contingency table of a term $t$ and a category $c$, where A is the number of times $t$ and $c$ co-occur, $B$ is the number of times $t$ occurs without $c$, $C$ is the number of times $c$ occurs without $t$, $D$ is the number of times neither $c$ nor $t$ occurs, and $N$ is the total number of documents, the term goodness measure is defined to be:

$$\chi^2(t,c) = \frac{N*(AD-CB)^2}{(A+C)*(B+D)*(A+B)*(C+D)}$$

The $\chi^2$ statistic has a natural value of zero if $t$ and $c$ are independent. We compute for each category the $\chi^2$ statistic between each unique term in a training dataset and that category, and then combine the category specific scores of each term into two scores:

$$\chi^2_{avg}(t) = \sum_{i=1}^{m} \Pr(c_i)\chi^2(t,c_i)$$

$$\chi^2_{max}(t) = \max_{i=1}^{m}\{\chi^2(t,c_i)\}$$

A major difference between $\chi^2$ and MI is that $\chi^2$ is a normalized value, and hence $\chi^2$ values are comparable across terms for the same category. However, this normalization breaks down if any cell in the contingency table is lightly populated, which is the case for low frequency terms. Therefore, the $\chi^2$ statistic is known not to be reliable for low frequency terms.

*Experimental Comparisons*

Various experimental comparisons of feature selection functions applied to TC contexts have been carried out (Yang and Pedersen 1997; Mladenic and Grobelnik 1998; Galavotti, Sebastiani et al. 2000; Caropreso, Matwin et al. 2001). In these experiments most functions have improved on the results for basic document frequency thresholding. For instance, (Yang and Pedersen 1997) have shown that, with various classifiers and various initial corpora, sophisticated techniques such as information gain can reduce the dimensionality of the term space by a factor of 100 with no loss (or even with a small increase) of effectiveness.

### 3.2.3  Combined Methods

Techniques for reducing the size of the dictionary are usually used in combination. It is very common first to eliminate stop words, then to convert the remaining terms to their stem roots. After that, rarely occurring terms are removed, by thresholding either their term frequency or their document frequency. Finally, a more elaborate method such as information gain can be used further to reduce the number of features.

## 3.3     Feature Vector Generation

The most popular text representation form is the *Vector Space Model* (Salton, Wong et al. 1975). The task of the vector generation step is to create a weighted vector $d = \left(w(d,t_1),...,w(d,t_m)\right)^T$ for any document $d$, where each weight $w(d,t_i)$ expresses the importance of term $t_i$ in document $d$.

After a vocabulary (list of all index terms appearing in the training documents) $V$ is built, it defines a $\|V\|$-dimensional vector space with the documents represented as vectors in this space. The value of the $j^{th}$ component of the $i^{th}$ vector is the weight of the $j^{th}$ index term in the $i^{th}$ document. The weights in the vector are determined by a weight function. Different weighting schemes exist, including:

- binary weighting has been used in (Apté, Damerau et al. 1994; Lewis and Ringuette 1994; Moulinier, Raskinis et al. 1996; Koller and Sahami 1997; Sebastiani, Sperduti et al. 2000) especially because of the symbolic, non-numeric nature of the learning systems
- term frequency
- TFIDF (McGill and Salton 1983)
- TFC (Salton and Buckley 1988)
- ITC (Buckley, Salton et al. 1995)
- and entropy weighting (Dumais 1991).

Some of the most commonly used methods are outlined below. Usually, the resulting vector is very sparse as most of the documents contain about 1 to 5% of the total number of terms in the vocabulary.

*Binary Representation*

One simple and common representation is the binary representation, where the appearance of an index is indicated with a 1 in the document vector representation. All non-present words have a weight of 0.

*Term Frequency*

Term frequency captures the number of occurrences of a term/index within a given document.

*TFIDF Representation*

TFIDF representation is an information-retrieval-style indexing technique that is widely used in text representation (a variant of it was used in Section 3.2.2).

$(TF\_IDF)_{ij} = TFij * IDF_i$ where $IDFi = \log_2 (n / DF_j)$
$TF_{ij}$ is the number of times term $t_j$ occurs in document $d_i$
$DF_j$ is the number of training documents in which word $t_j$ occurs at least once.
$n$ is the total number of training documents.

This weighting function encodes the intuitions that the more often a term occurs in a document, the more it is representative of the document's content, and that the more documents in which a term occurs, the less discriminating it is.

In order to make weights fall in the *[0,1]* interval and for documents to be represented by vectors of equal length, the weights resulting from the function *(TF_IDF)$_{ij}$* are normalized by 'cosine normalization', given by:

$$X_{ij} = \frac{\left(TF\_IDF\right)_{ij}}{\sqrt{\sum_{j=1}^{N}\left(TF\_IDF\right)_{ij}^{2}}}, \ 1 \leq i \leq n \text{ and } 1 \leq j \leq N$$

where *N* is the number of terms that occur at least once in the whole set of training documents.

## 4      Hypertext Categorization

It has been estimated that the World Wide Web comprises more than 9 billion pages (www.google.com). As long ago as 1998/9 it was estimated to be growing at a rate of 1.5 million pages a day (Bharat and Broder 1998; Lawrence and Giles 1999). Faced with such a huge volume of documents, search engines become limited: too much information to look at and too much information retrieved. The organization of web documents into categories will reduce the search space of search engines and improve their retrieval performance. A study by (Chen and Dumais 2000) showed that users prefer to navigate through directories of pre-classified content and that providing a categorized view of retrieved documents enables them to find more relevant information in a shorter time. The common use of the manually constructed category hierarchies for navigation support in Yahoo (www.yahoo.com) and other major web portals has also demonstrated the potential value of automating the process of hypertext categorization.

Text classification is a relatively mature area where many algorithms have been developed and many experiments conducted. For example, classification accuracy reached 87% (Chakrabarti, Dom et al. 1997) for some algorithms applied to known text categorization corpora (Reuters, 20-newsgroups etc.) where the vocabulary is coherent and the authorship is high. However, those same classifiers often perform badly on hypertext datasets. Hypertext classification poses new classification challenges in addition to those of text classification.

- Diversity of the web content: Web documents are diverse. They range from home pages, articles, tutorials etc. to portals.

- Diversity of authorship: the web is open to everybody. Millions of authors can put their scripts into the web. This leads to little consistency in the vocabulary.

- Sparse or non-existing text: Many web pages only contain a limited amount of text. In fact, many pages contain only images and no machine readable text at all.

The proverb "one picture is worth a thousand words" is widely applied by web authors.

Several variations of the classical classification algorithms have been developed to meet the particularities of the hypertext/text classification task.

As well as the above, automated hypertext categorization poses new research challenges because of the extra information in a hypertext document. Hyperlinks, HTML tags, metadata and linked neighbourhood all provide rich information for classifying hypertext that is not available in traditional text categorization.

Researchers have only recently begun to explore the issues of exploiting rich hypertext information for automated categorization. There is now a growing volume of research in the area of learning over web text documents. Since most of the documents considered are in HTML format, researchers have taken advantage of the structure of those pages in the learning process. The systems generated differ in performance because of the quantity and nature of the additional information considered.

(Benbrahim and Bramer 2004a) used the BankSearch dataset (Sinka, M. P. and D. W. Corne, 2002) to study the impact on classification of the use of metadata (page keywords and description), page title and link anchors in a web page. They concluded that the use of basic text content enhanced with weighted extra information (metadata + title + link anchors) improves the performance of three different classifiers. In (Benbrahim and Bramer 2004b), they used the same dataset to investigate the influence of the neighbourhood pages (incoming and outgoing pages of the target document) on classification accuracy. It was concluded that the intelligent use of this information helps improve the accuracy of the different classifiers used.

(Oh, Myaeng et al. 2000) reported some observations on a collection of online Korean encyclopaedia articles. They used system-predicted categories of the linked neighbours of a test document to reinforce the classification decision on that document and they obtained a 13% improvement over the baseline performance when using local text alone.

(Furnkranz 1999) used a set of web pages from the Web->KB corpus, created by the Carnegie-Mellon University World Wide Knowledge Base Project[2], to study the use of anchor text and the words near the anchor text in a web page to predict the class of the target page pointed to by the links. By representing the target page using the anchor words on all the links that point to it, plus the headlines that structurally precede the sections where links occur, the classification accuracy of a rule-learning system improved by 20%, compared with the baseline performance of the same system when using the local words in the target page instead.

(Slattery and Mitchell 2000) used the Web->KB university corpus, but studied alternative learning paradigms, namely, a First Order Inductive Learner which

---

[2]  http://www.cs.cmu.edu/~webkb/

exploits the relational structure among web pages, and a Hubs and Authorities style algorithm exploiting the hyperlink topology. They found that a combined use of these two algorithms performed better than using each alone.

(Yang, Slattery et al. 2002) have defined five hypertext regularities which may hold in a particular application domain, and whose presence may significantly influence the optimal design of a classifier. The experiments were carried out on 3 datasets and 3 learning algorithms. The results showed that the naïve use of the linked pages can be more harmful than helpful when the neighbourhood is noisy, and that the use of metadata when available improves classification accuracy.

(Attardi, Gulli et al. 1999) described an approach that exploits contextual information extracted from an analysis of the HTML structure of Web documents as well as the topology of the web. The results of the experiments with a categorization prototype tool were quite encouraging.

(Chakrabarti, Dom et al. 2002) studied the use of citations in the classification of IBM patents where the citations between documents were considered as hyperlinks, and the categories were defined on a topical hierarchy. Similar experiments on a small set of pages with real hyperlinks were also conducted. By using the system-predicted category labels for the linked neighbours of a test document to reinforce the category decision on that document, they obtained a 31% error reduction, compared with the baseline performance when using the linked documents, treating the words in the linked documents as if they were local.

## 5     Hypertext Representation

Before machine-learning methods can be applied to hypertext categorization, a decision first needs to be made about a suitable representation of HTML pages. Hypertext representation inherits all the basics and steps of the traditional text representation task described in Section 3. In addition, it takes advantage of the hidden information in HTML pages. Hyperlinks, HTML tags, metadata and information in a page's neighbours (documents pointing to and pointed to by the target page) are used to enrich the HTML pages' representation.

Vector Space Model (VSM) representations identify each document by a feature vector in a space in which each dimension corresponds to a distinct index term (feature). The set of features can be generated either manually or automatically based on a specific document collection, usually the set of training documents used as input for the learning algorithm. A given document vector has, in each component, a numerical value to indicate its importance. This value is expressed as a function of the frequency of the index term in the particular document. By varying this function, we can produce different term weightings. The resulting representation of the text is equivalent to the attribute-value representation, which is commonly used in machine learning.

There are several steps involved in transforming HTML-pages into feature vectors. Tokenization transforms a web-page into a sequence of word tokens that are used to build indexes during the indexation phase. There is an important difference between Hypertext Categorization and Text Categorization at the tokenization stage. In general, a hypertext dataset consists of a set of HTML source files, not plain text files. The aim of the tokenization step is the extraction of plain tokens (each a sequence of characters separated by one or more spaces) from each HTML page. After this, further normalization is applied to the set of tokens. Usually all letters are converted to lower-case as it is expected that case gives no information about the subject matter. Then HTML tags, scripts, punctuation, special characters and digits are removed. Tokens that contain numeric or non-alphanumeric characters are often omitted, even though sometimes they should not be. For example the token "c++" might indicate a programming language. Such a normalization step would ideally make use of both domain- and application-specific knowledge.

Following indexation, if training documents are being processed all distinct index terms are merged to generate a set of potential features that might be used in the dictionary. Usually, the large size of the dictionary is reduced at the dimensionality reduction step. If instead a new document is to be processed with an existing classifier and thus the dictionary is already built, the indexation task outputs only index terms which exist in the dictionary. Finally, a vector generation step calculates weights for all the index terms of any given HTML page.

Figures 2, 3 and 4 show an HTML document as we see it in an internet browser, part of the corresponding HTML source code and an example of what it might look like after pre-processing, respectively.

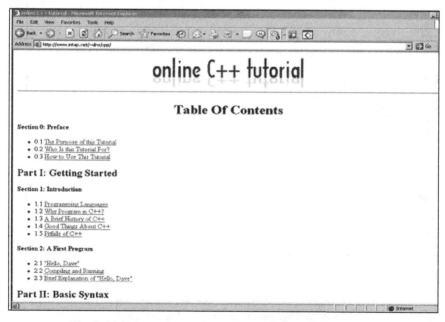

**Fig. 2.** An HTML Document as We See It in an Internet Browser

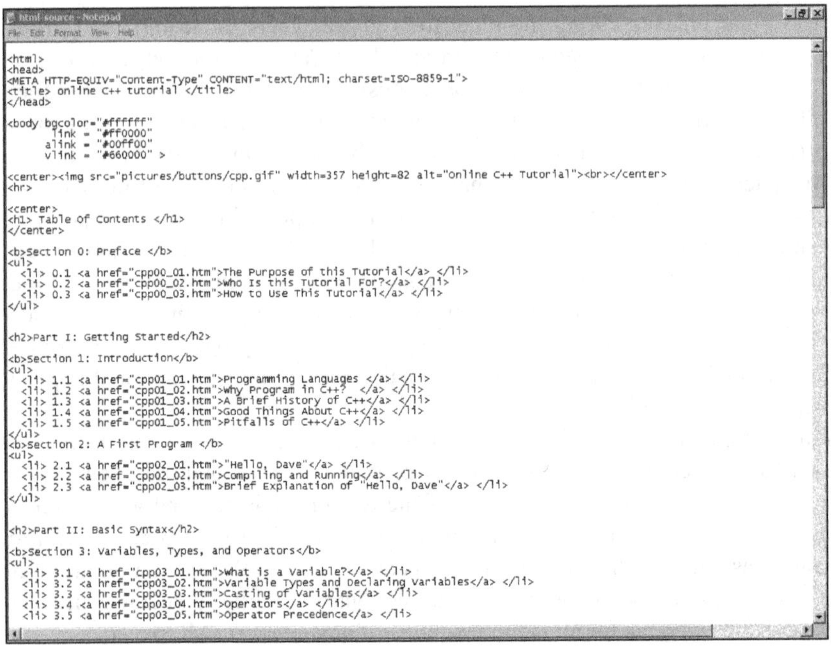

```
html source - Notepad
File Edit Format View Help

<html>
<head>
<META HTTP-EQUIV="Content-Type" CONTENT="text/html; charset=ISO-8859-1">
<title> online C++ tutorial </title>
</head>

<body bgcolor="#ffffff"
      link  = "#ff0000"
      alink = "#00ff00"
      vlink = "#660000" >

<center><img src="pictures/buttons/cpp.gif" width=357 height=82 alt="online C++ Tutorial"><br></center>
<hr>

<center>
<h1> Table of Contents </h1>
</center>

<b>Section 0: Preface </b>
<ul>
    <li> 0.1 <a href="cpp00_01.htm">The Purpose of this Tutorial</a> </li>
    <li> 0.2 <a href="cpp00_02.htm">who Is this Tutorial For?</a> </li>
    <li> 0.3 <a href="cpp00_03.htm">How to Use This Tutorial</a> </li>
</ul>

<h2>Part I: Getting Started</h2>

<b>Section 1: Introduction</b>
<ul>
    <li> 1.1 <a href="cpp01_01.htm">Programming Languages </a> </li>
    <li> 1.2 <a href="cpp01_02.htm">why Program in C++? </a> </li>
    <li> 1.3 <a href="cpp01_03.htm">A Brief History of C++</a> </li>
    <li> 1.4 <a href="cpp01_04.htm">Good Things About C++</a> </li>
    <li> 1.5 <a href="cpp01_05.htm">Pitfalls of C++</a> </li>
</ul>
<b>Section 2: A First Program </b>
<ul>
    <li> 2.1 <a href="cpp02_01.htm">"Hello, Dave"</a> </li>
    <li> 2.2 <a href="cpp02_02.htm">Compiling and Running</a> </li>
    <li> 2.3 <a href="cpp02_03.htm">Brief Explanation of "Hello, Dave"</a> </li>
</ul>

<h2>Part II: Basic Syntax</h2>

<b>Section 3: Variables, Types, and Operators</b>
<ul>
    <li> 3.1 <a href="cpp03_01.htm">what is a Variable?</a> </li>
    <li> 3.2 <a href="cpp03_02.htm">Variable Types and Declaring Variables</a> </li>
    <li> 3.3 <a href="cpp03_03.htm">Casting of Variables</a> </li>
    <li> 3.4 <a href="cpp03_04.htm">Operators</a> </li>
    <li> 3.5 <a href="cpp03_05.htm">Operator Precedence</a> </li>
```

**Fig. 3.** An Example Source Code of an HTML Document

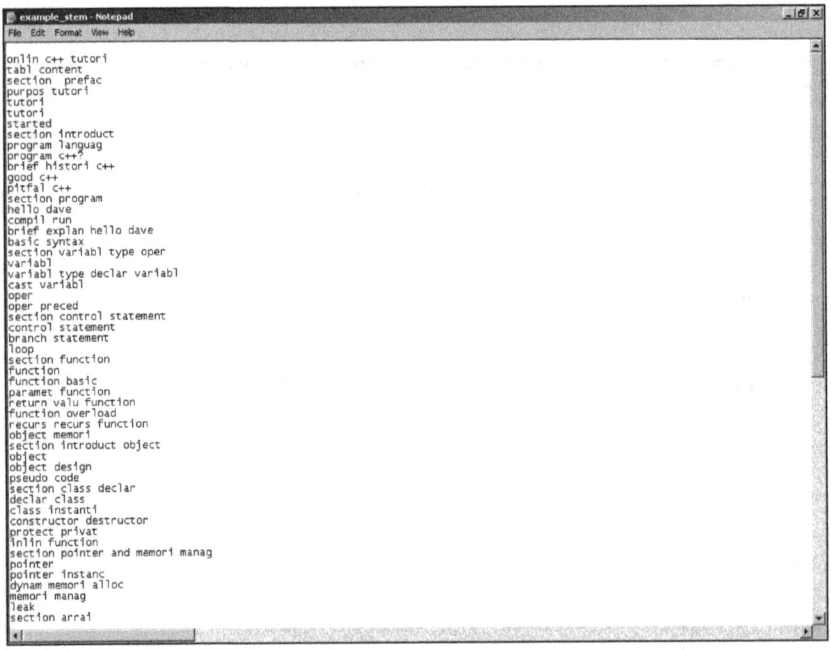

```
example_stem - Notepad
File Edit Format View Help

onlin c++ tutori
tabl content
section prefac
purpos tutori
tutori
tutori
started
section introduct
program languag
program c++?
brief histori c++
good c++
pitfal c++
section program
hello dave
compil run
brief explan hello dave
basic syntax
section variabl type oper
variabl
variabl type declar variabl
cast variabl
oper
oper preced
section control statement
control statement
branch statement
loop
section function
function
function basic
paramet function
return valu function
function overload
recurs recurs function
object memori
section introduct object
object
object design
pseudo code
section class declar
declar class
class instanti
constructor destructor
protect privat
inlin function
section pointer and memori manag
pointer
pointer instanc
dynam memori alloc
memori manag
leak
section arrai
```

**Fig. 4.** An Example Document after Pre-processing

## 5.1     Enriched Hypertext Representation

Web pages contain additional information such as metadata which it seems reasonable to suppose might be usable to improve categorization accuracy. This section addresses the open question of which extra information hidden in the HTML pages should be included, and how to include it. This affects both the tokenization and the vector generation phases.

An HTML document is much more than a simple text file. It is structured and connected with other HTML pages. The simplest way to represent a web page is to extract the text found in the BODY tag, but this representation does not exploit the rich information hidden in the hypertext document.

This extra information can be modelled with varying levels of detail, and might be divided into two major subsets. One subset includes information that is local to the HTML web document, mainly the information enclosed in the structure of the page itself such as the data in the META tag and TITLE tag, and the other contains the extra information present in the hyperlinks, i.e. information included in the page's neighbours (documents pointing to, and pointed to by the target page).

### 5.1.1   Which Extra Information?

*HTML Structure Data*

The basic structure of HTML pages has changed with HTML versions. At the beginning, HTML pages were quite simple, but as HTML has been extended, pages have taken a more subtle but still simple structure.

An ordinary HTML page has two main parts: the first line and the rest of the document (delimited by the **<HTML>** and **</HTML>** tags), which is also split into two other parts: the heading and the body of the page.

The very first line of an HTML page (<!DOCTYPE>) indicates to the browser which HTML version is used for this document.

The HTML page's heading, which follows immediately the <!DOCTYPE...> tag, is delimited by the <HEAD> and </HEAD> tags. This part contains the so called *meta* information, i.e. information about the content of the HTML page. For instance, one indicates information about the author of the page, the date of creation, keywords or a description of the content of the page, the page's title, and possibly other elements, like style sheets or JavaScript code. This part contains information that is not directly displayed in the browser.

```
<html>
<head>
<meta http-equiv="Content-Type" content="text/html; charset=UTF-8" />
<title>Home | University of Portsmouth</title>
<meta name="dc.keywords" content="University of Portsmouth, University, Portsmouth, degree, courses, research, postgraduate, undergraduate, study, studying" />
```

&lt;meta name="dc.description" content="Welcome to the University of Portsmouth, UK. Find out about studying in Portsmouth, our courses and research activities." /&gt; &lt;/head&gt;

The second part contains the content of the web page, i.e. what should be displayed by the web browser. This part is delimited by the &lt;BODY&gt; and &lt;/BODY&gt; tags. HTML tags (other than &lt;HEAD&gt;, &lt;BODY&gt;, &lt;HTML&gt; and &lt;META&gt;) are exclusively used in this part to format the document.

*Hyperlink Data*

Another source of extra information in HTML pages is the use of data in the linked neighbourhood (incoming and outgoing documents of the target page). This data can be put under two type levels, either local or remote hyperlink data.

The local hyperlink data (i.e. anchor data) can be captured from the content of the &lt;A&gt; tag in the target page. An example of its use:

&lt;A href = "./VB.html"&gt; Visual Basic tutorial &lt;/A&gt;.

In this example, the A tag is used to link the current page to another one. Users can have an idea about the content of the linked page by means of the content of the A tag, that is "Visual Basic tutorial". Usually, when a developer makes a link in his page, he tries to explain with few words the content of the linked page in the A tag. Therefore, it is assumed that the words used in this description are close to the subject of the linked page, and hence can be given special interest as they can represent the content of the linked page.

The remote hyperlink data might also be split into two subsets, namely, the set of links that the target page points to, and the set of links that point to the target page. To have access to this data, one should download the content of the linked pages.

### 5.1.2  How Much Extra Information?

Data in Meta-Description, Meta-Keywords and Title tags is generally considered more important and representative of a document's content than a word present in the BODY tag. Unfortunately, in a quantitative study conducted on a sample of web pages in (Pierre 2000), it was stated that although TITLE tags are common in HTML pages, their corresponding amount of text is relatively small with 89% of the titles containing only 1-10 words. Also, the titles often contain only names or terms such as ``home page", which are not particularly helpful for subject classification. On the other hand, Metatags for keywords and descriptions are found in only about a third of web sites. Most of the time these metatags contain between 11 and 50 words, with a smaller percentage containing more than 50 words (in contrast to the number of words in the body text which tended to contain more than 50 words).

There are other structured data in the BODY of HTML pages that might be taken into special consideration as well, such as the data in the heading tag, i.e. data enclosed in &lt;Hi&gt; and &lt;/Hi&gt;, or boldfaced data i.e. data between &lt;b&gt; and &lt;/b&gt;.

The use of links' content in hypertext representation might be straightforward, i.e. concatenate it to the target page. However, this blind use may harm the classification task, due to the fact that many pages point to or are pointed to by pages from different subjects, e.g., web pages of extremely diverse topics link to Yahoo! or BBC news web pages. Many techniques have been used to filter out the noisy link information. There are many different ways to measure how similar two documents are, including Euclidian distance, Cosine measure and Jaccard measure. The cosine measure is a very common similarity measure for textual datasets. It is given by:

$$\cos\_sim(D_A, D_B) = \frac{D_A . D_B}{\|D_A\|_2 . \|D_B\|_2}$$

$$= \frac{\sum_{i=1}^{n} w_{iA} . w_{iB}}{\sqrt{\sum_{i=1}^{n} w_{iA}^2} \sqrt{\sum_{i=1}^{n} w_{iB}^2}}$$

where $w_{iA}$ represents the weight of index $i$ in document $A$.

### 5.1.3 How to Include This Extra Information

*Structure Term Weighting*

To exploit the structural information in HTML pages, one should consider not only the frequency of the term in the document, but also the HTML tag the term is present in. The idea is to assign greater importance to terms that belong to tags that are believed to be more representative for web pages (META and TITLE tags). The term frequency of words might then be recomputed as an enhanced term frequency as follows:

$$ETF(t_i, D_j) = \sum_{tag_k} \left( w(tag_k) . TF(t_i, tag_k, D_j) \right)$$

where $tag_k$ is an HTML tag, $w(tag_k)$ is the weight we assign to $tag_k$ and $TF(t_i, tag_k, D_j)$ is the number of times the term $t_i$ is present in the tag $tag_k$ of the HTML document $D_j$.

$w(tag)$ can be defined as a function as:

$$w(tag) = \begin{cases} \alpha & \text{if tag} = \text{META} \\ \beta & \text{if tag} = \text{TITLE} \\ 1 & \text{if somewhere else} \end{cases}$$

*Linked Term Weighting*

There are many ways to incorporate the 'similar' linked neighbourhood in the target web document representation. The most straightforward method is to concatenate the content of the similar linked pages to the target web page. A new dictionary is then built to take into consideration the new indexes present in the neighbourhood, and that do not exist in the target pages.

A new enhanced term frequency scheme might be defined as:

$$EIF(t_i, D_j) = \sum_{source_k} \sum_{tag_k} \left( w_s(source_l) w_t(tag_k) . TF(t_i, tag_k, source_l, D_j) \right)$$

Where $tag_k$ is an HTML tag, $w_t(tag_k)$ is the weight we assign to $tag_k$, $source_l$ is the source of the term, i.e. term exists in target page, or a neighbour page, $w_s(source_l)$ is the weight we assign to $source_l$, and $TF(t_i, tag_k, source_l, D_j)$ is the number of times term $t_i$ is present in the tag $tag_k$ of the $source_l$ related to HTML document $D_j$.

$$w_t(tag) = \begin{cases} \alpha & \text{if tag} = \text{META} \\ \beta & \text{if tag} = \text{TITLE} \\ 1 & \text{if somewhere else} \end{cases}$$

$$w_s(source) = \begin{cases} \delta & \text{if source} = \text{incoming link} \\ \lambda & \text{if source} = \text{outgoing link} \\ 1 & \text{if source} = \text{target web page} \end{cases}$$

The drawback of this approach is that the resulting set of total features (size of the hypertext dictionary) is inflated.

The other approach is a variation of the previous one. In this case, no new dictionary is built, i.e. just terms already existing in the original HTML page are considered, and no extra terms from the neighbourhood are added. The new term frequencies are recomputed using the same above formula.

Another approach is to distinguish the terms absorbed from the neighbours from the original local terms by adding a prefix, so that the classifier can distinguish between the local and non-local terms. The disadvantage of this method is the term is split into many forms which can make it relatively rare, even if the corresponding original term is not that rare. It also explodes the size of the new dictionary, and hence, the classifier faces a challenge as the number of features increases while the number of documents remains the same.

## 5.2     Experiments with FSS-SVM

With the techniques described above, it is possible to map any HTML document into its vector representation, which is then suitable as input for any of a range of standard learning algorithms, such as decision trees (Quinlan, 1986) and support vector machines (Vapnik, 1995). A recent example of the use of a new form of Fuzzy Semi-supervised Support Vector Machine (FSS-SVM) to classify web pages with "enriched' representation is given in (Benbrahim and Bramer, 2008).

In all experiments, the fuzzy semi-supervised support machine performed better than its supervised version when the number of labelled training documents is small, i.e. FSS-SVM can achieve a specific level of classification accuracy with much less labelled training data. For example, with only 550 labelled training examples for the BankSearch dataset (50 documents per class), FSS-SVM reached 65% classification accuracy, using the F1 measure of accuracy (Bramer, 2007), while the traditional SVM classifier achieved only 50%. To reach 0.65 classification accuracy, SVM required about 1100 labelled training documents and FSS-SVM only 550.

For the Web->KB dataset, the performance increase was also substantial. For 80 labelled training examples (20 documents per class), SVM obtained 29% accuracy (F1 measure) and FSS-SVM 59%, reducing classification error by 30%. (Benbrahim and Bramer, 2008) also reports other experimental findings and gives full details of the fuzzy semi-supervised support vector machine algorithm and the enriched representation of hypertext documents used.

# References

Amati, G., Crestani, F.: Probabilistic learning for selective dissemination of information. Information Processing and Management 35(5), 633–654 (1999)

Andrews, K.: The development of a fast conflation algorithm for English. Dissertation submitted for the Diploma in Computer Science (unpublished), University of Cambridge (1971)

Apté, C., Damerau, F., et al.: Automated learning of decision rules for text categorization. ACM Transactions on Information Systems (TOIS) 12(3), 233–251 (1994)

Attardi, G., Gulli, A., et al.: Automatic Web Page Categorization by Link and Context Analysis. In: Proceedings of THAI'99, pp. 105–119 (1999)

Baker, L.D., McCallum, A.K.: Distributional clustering of words for text classification. In: Proceedings of the 21st annual international ACM SIGIR conference on Research and development in information retrieval, pp. 96–103 (1998)

Benbrahim, H., Bramer, M.: An empirical study for hypertext categorization. In: IEEE International Conference on Systems, Man and Cybernetics, 2004, vol. 6 (2004a)

Benbrahim, H., Bramer, M.: Neighbourhood Exploitation in Hypertext Categorization. In: Proceedings of the Twenty-fourth SGAI International Conference on Innovative Techniques and Applications of Artificial Intelligence, Cambridge, December 2004, pp. 258–268 (2004b)

Benbrahim, H., Bramer, M.: A Fuzzy Semi-Supervised Support Vector Machines Approach to Hypertext Categorization. In: Artificial Intelligence in Theory and Practice II, pp. 97–106. Springer, Heidelberg (2008)

Bharat, K., Broder, A.Z.: A Technique for Measuring the Relative Size and Overlap of Public Web Search Engines. WWW7 / Computer Networks 30(1-7), 379–388 (1998)

Borko, H., Bernick, M.: Automatic Document Classification. Journal of the ACM (JACM) 10(2), 151–162 (1963)

Bramer, M.A.: Principles of Data Mining. Springer, Heidelberg (2007)

Buckley, C., Salton, G., et al.: Automatic query expansion using SMART. In: TREC 3, Overview of the Third Text Retrieval Conference (TREC-3), pp. 500–225 (1995)

Caropreso, M.F., Matwin, S., et al.: A learner-independent evaluation of the usefulness of statistical phrases for automated text categorization. In: Text Databases and Document Management: Theory and Practice, pp. 78–102 (2001)

Cavnar, W.B., Trenkle, J.M.: N-Gram based document categorization. In: Proceedings of the Third Symposium on Document Analysis and Information Retrieval, Las Vegas, pp. 161–176 (1994)

Chakrabarti, S., Dom, B., et al.: Using taxonomy, discriminants, and signatures for navigating in text databases. In: Proceedings of the 23rd VLDB Conference, pp. 446–455 (1997)

Chakrabarti, S., Dom, B., et al.: Scalable feature selection, classification and signature generation for organizing large text databases into hierarchical topic taxonomies. The VLDB Journal The International Journal on Very Large Data Bases 7(3), 163–178 (1998)

Chakrabarti, S., Dom, B.E., et al.: Enhanced hypertext categorization using hyperlinks. Google Patents (2002)

Chen, H., Dumais, S.: Bringing order to the web: automatically categorizing search results. In: Proceedings of the SIGCHI conference on Human factors in computing systems, pp. 145–152 (2000)

Clack, C., Farringdon, J., et al.: Autonomous document classification for business. In: Proceedings of the 1st International Conference on Autonomous Agents, pp. 201–208 (1997)

Cohen, W.W., Hirsh, H.: Joins that generalize: text classification using Whirl. In: Proceedings of KDD-98, 4th International Conference on Knowledge Discovery and Data Mining, pp. 169–173 (1998)

Cohen, W.W., Singer, Y.: Context-Sensitive Learning Methods for Text Categorization. In: Conference on Research and Development in Information Retrieval (SIGIR), pp. 307–315 (1998)

Creecy, R.H.: Trading MIPS and Memory for Knowledge Engineering: Automatic Classification of Census Returns on a Massively Parallel Supercomputer. Thinking Machines Corp. (1991)

Dagan, I., Karov, Y., et al.: Mistake-driven learning in text categorization. In: Proceedings of the Second Conference on Empirical Methods in NLP, pp. 55–63 (1997)

Dattola, R.T.: FIRST: Flexible Information Retrieval System for Text. J. Am. Soc. Inf. Sci. 30(1) (1979)

De Heer, T.: The application of the concept of homeosemy to natural language information retrieval. Information Processing & Management 18(5), 229–236 (1982)

Deerwester, S., Dumais, S.T., et al.: Indexing by latent semantic analysis. Journal of the American Society for Information Science 41(6), 391–407 (1990)

Domingos, P., Pazzani, M.: On the Optimality of the Simple Bayesian Classifier under Zero-One Loss. Machine Learning 29(2), 103–130 (1997)

Dumais, S., Platt, J., et al.: Inductive learning algorithms and representations for text categorization. In: Proceedings of the seventh international conference on Information and knowledge management, pp. 148–155 (1998)

Dumais, S.T.: Improving the retrieval of information from external sources. Behavior Research Methods, Instruments and Computers 23(2), 229–236 (1991)

Escudero, G., Marquez, L., et al.: Boosting Applied to Word Sense Disambiguation. Arxiv preprint cs.CL/0007010 (2000)

Field, B.: Towards automatic indexing: automatic assignment of controlled-language indexing and classification from free indexing. Journal of Documentation 31(4), 246–265 (1975)

Fuhr, N., Buckley, C.: A probabilistic learning approach for document indexing. ACM Transactions on Information Systems (TOIS) 9(3), 223–248 (1991)

Furnkranz, J.: Exploiting structural information for text classification on the WWW. In: Hand, D.J., Kok, J.N., R. Berthold, M. (eds.) IDA 1999. LNCS, vol. 1642, pp. 487–497. Springer, Heidelberg (1999)

Galavotti, L., Sebastiani, F., Simi, M.: Experiments on the use of feature selection and negative evidence in automated text categorization. In: Borbinha, J.L., Baker, T. (eds.) ECDL 2000. LNCS, vol. 1923, pp. 59–68. Springer, Heidelberg (2000)

Gale, W.A., Church, K.W., et al.: A method for disambiguating word senses in a large corpus. Computers and the Humanities 26(5), 415–439 (1992)

Gray, W.A., Harley, A.J.: Computer-assisted indexing. Inform. Storage Retrieval 7(4), 167–174 (1971)

Hersh, W., Buckley, C., et al.: OHSUMED: An interactive retrieval evaluation and new large test collection for research. In: Proceedings of the 17th annual international ACM SIGIR conference on Research and development in information retrieval, pp. 192–201 (1994)

Hull, D.: Improving text retrieval for the routing problem using latent semantic indexing. In: Proceedings of the 17th annual international ACM SIGIR conference on Research and development in information retrieval, pp. 282–291 (1994)

Ittner, D.J., Lewis, D.D., et al.: Text categorization of low quality images. In: Symposium on Document Analysis and Information Retrieval, pp. 301–315 (1995)

Iyer, R.D., Lewis, D.D., et al.: Boosting for document routing. In: Proceedings of the ninth international conference on Information and knowledge management, pp. 70–77 (2000)

Joachims, T.: A Probabilistic Analysis of the Rocchio Algorithm with TFIDF for Text Categorization, School of Computer Science, Carnegie Mellon University (1996)

Joachims, T.: Text Categorization with Suport Vector Machines: Learning with Many Relevant Features. Springer, London (1998)

Kim, Y.H., Hahn, S.Y., et al.: Text filtering by boosting naive Bayes classifiers. In: Proceedings of the 23rd annual international ACM SIGIR conference on Research and development in information retrieval, pp. 168–175 (2000)

Koller, D., Sahami, M.: Hierarchically classifying documents using very few words. In: Proceedings of the Fourteenth International Conference on Machine Learning, pp. 170–178 (1997)

Lam, S.L.Y., Lee, D.L.: Feature reduction for neural network based text categorization. In: Proceedings of 6th International Conference on Database Systems for Advanced Applications, pp. 195–202 (1999)

Lam, W., Ho, C.Y.: Using a generalized instance set for automatic text categorization. In: Proceedings of the 21st annual international ACM SIGIR conference on Research and development in information retrieval, pp. 81–89 (1998)

Lang, K.: Newsweeder: Learning to filter netnews. In: Proceedings of the Twelfth International Conference on Machine Learning, pp. 331–339 (1995)

Larkey, L.S.: Automatic essay grading using text categorization techniques. In: Proceedings of the 21st annual international ACM SIGIR conference on Research and development in information retrieval, pp. 90–95 (1998)

Larkey, L.S.: A patent search and classification system. In: Proceedings of the fourth ACM conference on Digital libraries, pp. 179–187 (1999)

Larkey, L.S., Croft, W.B.: Combining classifiers in text categorization. In: Proceedings of the 19th annual international ACM SIGIR conference on Research and development in information retrieval, pp. 289–297 (1996)

Lawrence, S., Giles, C.L.: Accessibility of information on the web. Nature 400, 107 (1999)

Lewis, D.D.: An evaluation of phrasal and clustered representations on a text categorization task. In: Proceedings of the 15th annual international ACM SIGIR conference on Research and development in information retrieval, pp. 37–50 (1992)

Lewis, D.D.: Feature selection and feature extraction for text categorization. In: Proceedings of the workshop on Speech and Natural Language, pp. 212–217 (1992)

Lewis, D.D.: Representation and learning in information retrieval. PhD Thesis, Department of Computer and Information Science, University of Massachusetts (1992)

Lewis, D.D.: Naive (Bayes) at forty: The independence assumption in information retrieval. In: Nédellec, C., Rouveirol, C. (eds.) ECML 1998. LNCS, vol. 1398, pp. 4–15. Springer, Heidelberg (1998)

Lewis, D.D., Ringuette, M.: A comparison of two learning algorithms for text categorization. In: Third Annual Symposium on Document Analysis and Information Retrieval, pp. 81–93 (1994)

Li, H., Yamanishi, K.: Text classification using ESC-based stochastic decision lists. In: Proceedings of the eighth international conference on Information and knowledge management, pp. 122–130 (1999)

Li, Y.H., Jain, A.K.: Classification of Text Documents. The Computer Journal 41(8), 537 (1998)

Lovins, J.B.: Development of a Stemming Algorithm. MIT Information Processing Group, Electronic Systems Laboratory (1968)

Luhn, H.P.: The automatic creation of literature abstracts. IBM Journal of Research and Development 2(2), 159–165 (1958)

Manning, C.D., Schütze, H.: Foundations of Statistical Natural Language Processing. MIT Press, Cambridge (1999)

Masand, B.: Optimizing confidence of text classification by evolution of symbolic expressions. Mit Press In Series In Complex Adaptive Systems, pp. 445–458 (1994)

Masand, B., Linoff, G., et al.: Classifying news stories using memory based reasoning. In: Proceedings of the 15th annual international ACM SIGIR conference on Research and development in information retrieval, pp. 59–65 (1992)

McCallum, A., Nigam, K.: Employing EM in pool-based active learning for text classification. In: Proceedings of ICML-98, 15th International Conference on Machine Learning, pp. 350–358 (1998)

McGill, M.J., Salton, G.: Introduction to modern information retrieval. McGraw-Hill, New York (1983)

Miller, G., Princeton, U., et al.: WordNet. MIT Press, Cambridge (1998)

Mladenic, D., Grobelnik, M.: Word sequences as features in text-learning. In: Proceedings of ERK-98, the Seventh Electrotechnical and Computer Science Conference, pp. 145–148 (1998)

Moulinier, I., Ganascia, J.G.: Applying an existing machine learning algorithm to text categorization. In: Wermter, S., Scheler, G., Riloff, E. (eds.) IJCAI-WS 1995. LNCS, vol. 1040, pp. 343–354. Springer, Heidelberg (1996)

Moulinier, I., Raskinis, G., et al.: Text categorization: a symbolic approach. In: Proceedings of the Fifth Annual Symposium on Document Analysis and Information Retrieval (1996)

Ng, H.T., Goh, W.B., et al.: Feature selection, perception learning, and a usability case study for text categorization. In: Proceedings of the 20th annual international ACM SIGIR conference on Research and development in information retrieval, pp. 67–73 (1997)

Nigam, K., Ghani, R.: Analyzing the effectiveness and applicability of co-training. In: Proceedings of the ninth international conference on Information and knowledge management, pp. 86–93 (2000)

Oh, H.J., Myaeng, S.H., et al.: A practical hypertext catergorization method using links and incrementally available class information. In: Proceedings of the 23rd annual international ACM SIGIR conference on Research and development in information retrieval, pp. 264–271 (2000)

Petrarca, A.E., Lay, W.M.: Use of an automatically generated authority list to eliminate scattering caused by some singular and plural main index terms. Proceedings of the American Society for Information Science 6, 277–282 (1969)

Pierre, J.M.: Practical Issues for Automated Categorization of Web Sites. In: Electronic Proc. ECDL 2000 Workshop on Semantic Web (2000)

Porter, M.: An Algorithm for Suffix Stripping Program. Program 14(3), 130–137 (1980)

Quinlan, J.R.: Induction of decision trees. Machine Learning 1(1), 81–106 (1986)

Ruiz, M.E., Srinivasan, P.: Hierarchical neural networks for text categorization (poster abstract). In: Proceedings of the 22nd annual international ACM SIGIR conference on Research and development in information retrieval, pp. 281–282 (1999)

Sable, C.L., Hatzivassiloglou, V.: Text-based approaches for non-topical image categorization. International Journal on Digital Libraries 3(3), 261–275 (2000)

Salton, G., Buckley, C.: Term-weighting approaches in automatic text retrieval. Information Processing and Management 24(5), 513–523 (1988)

Salton, G., Wong, A., et al.: A vector space model for information retrieval. Communications of the ACM 18(11), 613–620 (1975)

Schapire, R.E., Singer, Y.: BoosTexter: A Boosting-based System for Text Categorization. Machine Learning 39(2), 135–168 (2000)

Schütze, H., Hull, D.A., et al.: A comparison of classifiers and document representations for the routing problem. In: Proceedings of the 18th annual international ACM SIGIR conference on Research and development in information retrieval, pp. 229–237 (1995)

Sebastiani, F., Sperduti, A., et al.: An improved boosting algorithm and its application to automated text categorization (2000)

Sinka, M.P., Corne, D.W.: A large benchmark dataset for web document clustering. Soft Computing Systems: Design, Management and Applications 87, 881–890 (2002)

Sj, C., Waltz, D.J.: Trading mips and memory for knowledge engeneering. Communications of the ACM 35, 48–64 (1992)

Slattery, S., Mitchell, T.: Discovering test set regularities in relational domains. In: Proc. ICML (2000)

Slonim, N., Tishby, N.: The power of word clusters for text classification. In: Proceedings of ECIR-01, 23rd European Colloquium on Information Retrieval Research (2001)

Taira, H., Haruno, M.: Feature selection in SVM text categorization. In: Proceedings of the sixteenth national conference on Artificial intelligence and the eleventh Innovative applications of artificial intelligence conference innovative applications of artificial intelligence table of contents, pp. 480–486 (1999)

Tauritz, D.R., Kok, J.N., et al.: Adaptive Information Filtering using evolutionary computation. Information Sciences 122(2-4), 121–140 (2000)

Tzeras, K., Hartmann, S.: Automatic indexing based on Bayesian inference networks. In: Proceedings of the 16th annual international ACM SIGIR conference on Research and development in information retrieval, pp. 22–35 (1993)

Vapnik, V.N.: The Nature of Statistical Learning Theory. Springer, New York (1995)

Wai, L.A.M., Fan, L.: Using a Bayesian Network Induction Approach for Text Categorization. In: Proceedings of the 15th International Joint Conference on Artificial Intelligence, pp. 745–750 (1997)

Weiss, S.M., Apte, C., et al.: Maximizing text-mining performance. IEEE Intelligent Systems 14(4), 63–69 (1999)

Wiener, E., Pedersen, J.O., et al.: A neural network approach to topic spotting. In: Proceedings of the Fourth Annual Symposium on Document Analysis and Information Retrieval (SDAIR'95), pp. 317–332 (1995)

Yang, Y., Chute, C.G.: An example-based mapping method for text categorization and retrieval. ACM Transactions on Information Systems (TOIS) 12(3), 252–277 (1994)

Yang, Y., Liu, X.: A re-examination of text categorization methods. In: Proceedings of the 22nd annual international ACM SIGIR conference on Research and development in information retrieval, pp. 42–49 (1999)

Yang, Y., Pedersen, J.O.: A comparative study on feature selection in text categorization. In: Proceedings of the Fourteenth International Conference on Machine Learning 97 (1997)

Yang, Y., Slattery, S., et al.: A Study of Approaches to Hypertext Categorization. Journal of Intelligent Information Systems 18(2), 219–241 (2002)

# Future Challenges for Autonomous Systems

Helder Coelho

University of Lisbon, Campo Grande, 1749-066 Lisboa, Portugal
hcoelho@di.fc.ul.pt

**Abstract.** The domain of intelligent creatures, systems and entities is suffering today profound changes, and the pace of more than a hundred meetings (congresses, conferences, workshops) per year shows there is a very large community of interest, eager of innovations and creativity. There is now no unanimity and homogeneity of the crowd, no convergence on what concerns scientific or technological goals, and recent surveys offer us strange results about the desires of industry and academy. However, observing recent conferences, we can work out some tendencies and move toward the future, yet conflicts are present concerning the aims the multiple communities pursue because some themes are relevant for several communities. Also, interleaving of areas generates points of friction between what must be done next.

## 1    Introduction

The NASA space programme is a good example for checking why autonomous systems became important and an aid to achieve complex aims. The Deep Space One mission started in 1999 (see also the real-time fault diagnosis system on the space shuttle by Georgeff and Ingrand) to validate technologies (e.g. autonomy), to support non-pre-planned goals, to make tests with simulated failures, and to open new avenues for more advanced applications. Examples of human-level intelligence are made today with software agents, surrounded by complex environments, a completely different kind than the single-task or smaller-scale agents of yesterday.

An agent is the sum of a system's knowledge (represented with particular constructs) and the processes that operate on those constructs. The agent research field, within computing sciences, aims to produce more robust and intuitive artifacts, with the ability to deal with the unexpected. The building of these systems has turned out to be more complex than desired. This has pushed the whole community to refocus on the problem of the construction of autonomous and multi-agent systems (MAS) by adopting less traditional software engineering principles, such as agent oriented programming languages and advanced software frameworks. Usually an agent can carry out several tasks simultaneously. In JADE (Java Agent DEvelopment framework) the tasks that an agent

M. Bramer (Ed.): Artificial Intelligence, LNAI 5640, pp. 39–52, 2009.

performs are encapsulated in a class called behaviour. This class contains an action method which executes the task and as it can be executed several times there is another method to decide when this loop ends. An agent may contain several behaviours which are scheduled in a cooperative way, using a round-robin cyclic algorithm. At each instant, the scheduler picks a behaviour and executes its action method. If the behaviour ends, it is removed from the agent´s list of behaviours. Otherwise, the scheduler picks another behaviour and repeats the same process. In the past, all behaviours in JADE were scheduled with the same priority, so the programmer could not assign different priorities to behaviours to give them different processor times. However, there are situations in which the use of priorities is needed in order to achieve a desired functioning of MAS. Two new behaviours that allow the use of priorities in JADE were implemented recently by (Suárez-Romero et al, 2006).

MAS are seen today as an appropriate technology to develop complex distributed software systems. The development of such systems requires an agent-oriented methodology that guides developers to build and maintain such complex entities. From the software engineering point of view, the systematic development of software should follow several phases. The number and purpose of these phases depend on the problem domain and the methodology. It is easy to find requirement elicitation, system analysis, architectural design, implementation, and validation/verification stages in most software methodologies, but the main focus of most of the agent-oriented methodologies (GAIA, Prometheus, Tropos, MaSE, SODA, MAS-CommonKADS, INGENIAS) has been on the analysis and design phases, paying less attention to implementation and testing. So, efforts have been made on languages for MAS and platforms and tools for multi-agent development. Yet, more focus must be given to individual agents, their organisations and coordination mechanisms, and shared environments, and this direction of work implies a more mature methodology with guidelines, concepts, methods and tools to facilitate the development. As a consequence, a unifying perspective of design, implementation and verification is desirable.

The world has changed. Modern systems are now very different from traditional information systems. They became open, made with autonomous, heterogeneous parts interacting dynamically. On account of this, some of the old software techniques were forced to leave the scene and to allow agents, adaptive reasoning and interactions to take the stage (Wegner, 1997).

The field of autonomous agents and multi-agent systems (AAMAS), or agents worlds, covers paradigms for software engineering and tools for understanding human societies. It contains also issues of distributed and concurrent systems, artificial intelligence, economics, game theory and social sciences. This is a large space of interactions, an inter- and multidisciplinary space.

Moreover, in recent years, task and team work were recognized as separable. This conviction supported the inquiry around a particular task and the interactions among agents, and pushed the need for a central mechanism for controlling the behaviour of agents in open environments. Autonomic computing became an important research topic aimed at answering the needs of a new range of problem domains, electronic institutions (eAdministration and eGovernment) in web communities, online markets, patient care-delivery, virtual reality and simulation. Open worlds increased in number and complexity challenging agent architects to facilitate automated decision making by all sorts of agents, including robots.

As a matter of fact, throughout the 90s we observed a move from algorithms to interactive computation because there was an imperative shift from mainframes to networks, wireless devices and intelligent appliances, and, at the same time, from number crunching to embedded systems and graphical user interfaces, and, also, from procedure-oriented to object-based and distributed programming.

New (virtual) information spaces built around digital cities, a kind of web portal (e.g. PortEdu), combine social, political and economic activities, including online forms, (profit and non-profit) data services shopping, education (e.g. AMPLIA environment for Medicine) and entertainment. Government regulations extended laws with guidance to corporate and public actions, requested norms (obligations, prohibitions and permissions) for disciplining social interactions, and attracting advertisers and businesses among private and public companies. City informatization emerged as government initiatives and data services on accumulating urban information were requested. Otherwise, they recognize the relevance of usability, they are well maintained, use proprietary software and rely on powerful search engines.

Institutions have shown they are a mechanism to make agent interactions more effective, structured, coordinated and efficient. New tools were created to assist in the design and verification processes. For example, landmarks play a potential role in building a bridge between the rigidity of the protocols and the flexibility of norms for regulating the behavior of autonomous agents.

In negotiation, everybody tries to reach a profitable result. But, the outcome of future business, or the success of social turns, depends heavily on the relationships among agents, the trust they have on others and their own good reputation. Trust and reputation models, constructed with values from past interactions with the environment (and extra information) now influence the agent's decision-making process, even facilitate dealing with uncertain data.

In connected communities, software entities act on behalf of users and cooperate with infocitizens. Therefore, issues related to social and organizational aspects of agency are of keen importance, specially the questions of individual power, associated with decision-taking and reasoning (management of organizations, enforcement of norms), and of (adjustable) autonomy of agents in crisis situations.

In the following, we start by enumerating the current positive and negative aspects, observe the state of the art of autonomous agents and multi-agent systems, and go deep into the field to discover emerging trends and future challenges.

## 2     Today's Potentialities and Difficulties

The history of computing is marked by five ongoing trends: ubiquity, interconnection, intelligence, delegation and human-orientation, which are all associated with agents, a sort of computer entity capable of flexible, autonomous (problem-solving) action, situated in dynamic, open, unpredictable and distributed domains. Agents (synthetic and physical) are currently viewed as a metaphor for the design of complex distributed computational systems, as a source of technologies, and as simulation models of complex real-world systems, such as in biology or economics. In what concerns design, agents are mixed today with agent-oriented software engineering, architectures, mobile agents, infrastructures, and electronic institutions. The agent technologies are diverse, such as planning, communication languages, coordination mechanisms, matchmaking architectures, information agents (and basic ontologies), auction mechanism design, negotiation strategies, and learning.

There is a close connection between MAS and software architectures: a multi-agent system provides the software to solve the problem by structuring the system as a number of interacting autonomous entities embedded in an environment in order to achieve the functional and quality requirements of the system, and a software architecture is considered as the structure of a system which comprises software elements and the relationships among the elements. So, MAS are a valuable way to solve software intricacies in a large range of possible directions to face problems. Typical architectural elements of a multi-agent system are agents, environments, resources, or services. The relationships between those elements are very diverse, ranging from environment mediated interaction between cooperative agents via virtual trails to complex negotiation protocols in a society of self-interested agents. This analogy suggests the integration of MAS with mainstream software engineering. Meanwhile current practice considers it as a radically new way of doing software.

MAS engineering has not yet addressed the issues of autonomy and interoperability in depth because FIPA-ACL infrastructures have adopted ad-hoc and developer-private communications assumptions, based upon reasons of communication efficiency or developer convenience. So the problem of interaction exists (agents are often hand-crafted) and requires new directions apart from those already adopted by the tools Jadex and Jason.

In the aftermath of a large scale disaster, agents' decisions range from individual (e.g. survival) to collective (e.g. victims' rescue or fire extinction) attitudes, thus shaping a 2-strata decision model (Silva and Coelho, 2007). However, current decision-theoretic models are either purely individual or purely collective and find it difficult to deal with motivational attitudes. On the other hand, mental-state based models find it difficult to deal with uncertainty. A way out is the making of hybrid decision models: i) the collective versus individual (CvI), which integrates both strata quantitative evaluation of decision making, and ii) the CvI-JI which extends the CvI model, using the joint-intentions formulation of teamwork, to deal with collective mental-state motivational attitudes. Both models have been evaluated from an experimental and case study based outlook that explores the tradeoff between cost reduction and loss of optimality while learning coordination skills in a partially observable stochastic domain.

At the end of the 80s, agents were capable of playing the role of intentional systems, due to the Bratman's BDI (Belief-Desire-Intention) model, based upon Folk Psychology where behaviour is predicted and explained through the attribution of attitudes, such as believing, desiring, wanting, hoping, etc. Several BDI architectures (PRS, IRMA, dMARS) were implemented and explored in real problem-domains where issues such as organizations, interaction, cooperation, coordination, communication, competition or negotiation are the most interesting features.

The industrial strength of software is marked by a fundamental obstacle to take-up, the lack of mature agent software methodology, where joint goals, plans, norms and protocols regularly support coordination and interaction in organizations, and libraries of agent and organization models, of communication languages and patterns, or ontology patterns are available.

Over the past 20 years of research, logical theories of intelligent (rational) agents have been improved and refined and have served as the basis for executable specifications in order to implement diverse behaviours. Although this provides strong correctness, it became clear that those idealized specifications are inappropriate in practical situations because an agent has many resource-bounds (time and memory) to contend with. A way out is to represent and execute resource-bounded agents, and the reason justifying such a framework is the restriction of the amount of reasoning (temporal and doxastic) that the agent is allowed. For example, good candidates are the logic ATLBM (Alternating-time Temporal Logic with Bounded Memory), of Agotnes and Walther, able to describe the strategic abilities (decisions for certain circumstances) of coalitions of agents with bounded memory, imperfect recall and incomplete information (future inquiry: logical properties (complete axiomatizations) and computational complexity of the different logics), and the logic of situated-resource-bounded agents of Alechina and Logan.

The gap between realization (e.g. with BDI logic) and implementation (with AgentSpeak language) of intelligent agents remained unclear till today because both sides are superficially related: BDI notions in agent logics, modelled by abstract relations in modal logic, are not grounded in agent computations, and have no association with real behaviours. The current work of Meyer's group at Utrecht University is around the quest for the holy grail to agent verification and intended to decrease the gap between the formal aspects and the realization/implementation of intelligent agents.

# 3     Technologies, Tools and Techniques

Theoretical and practical issues associated with the developing and deploying of multi-agent systems are now discussed within the interdisciplinary space of formal methods, programming languages, methodologies, techniques and principles (e.g.

LADS´007 Workshop within the federated set of Workshops MALLOW). Along formal methods, current work includes the integration of two different concepts of belief (either as a probability distribution or as a logical formula) of Lloyd and Ng. Another good example of methodologies, in progress, is the MaSE for the development of embedded real-time systems of Badr, Mubarak and Göhner, well suited to aircrafts and industrial process controllers. Regarding tools and techniques, a modern example is associated with measuring the complexity of simulations of Klügl, because the variety of multi-agent applications is very large, in particular when compared with traditional simulation paradigms. The actual complexity of models is hidden, and there is a need to characterize it by introducing metrics for the properties of simulations. Finally, in what concerns programming languages, a nice example is the one on BDI agent organization by Hepple, Dennis and Fisher, because the BDI model style is common and there is agreement on the core functionality of agents. This research line aims at a unifying framework for the core aspects of agent organization, including groups, teams, roles and organizations as well. The purpose was to define a simple organizational mechanism, derived from the METATEM programming language, and show how several well known approaches can be embedded within it (the mechanism is intended to be independent of the underlying agent language).

Multi-agent systems (MAS) became one of the key technologies for software development today. This was achieved because the formal aspects of agency were deeply studied in a multi-disciplinary way with the help of logic and theoretical computer science (e.g. formal methods for verification and model checking, cooperation, planning, communication, coordination, negotiation, games, and reasoning under uncertainty in a distributed environment). Attacking agent organization is possible by the construction of a bridge between organizational theory and logics for social concepts governing MAS in open environments, as Virginia and Frank Dignum are doing. They are formalizing group organization capabilities and responsibilities, and relating two kinds of organizational structure, such as hierarchies and networks. Lorini, Herzig, Broersen and Trioquard are grounding power on actions and mental attitudes (beliefs, intentions) via Intentional Agency Logic.

MAS can be observed from the perspective of the Population-Organizational model, a minimal semantic model where the performance of organizational roles by agents and the realization of organizational links by social exchanges between agents are the key mechanisms for the implementation of an organization structure by a population structure. The structural dynamics of a MAS may then be modelled as a set of transformations on the system´s overall population-organization structure. Rocha Costa and Dimuro illustrated this approach by introducing a small set of operational rules for an exchange value based dynamics of organizational links.

In a case study of natural resources management, Adamatti, Sichman and Coelho proposed a new software architecture, called ViP-GMABS, which enabled virtual players to be associated in the GMABS methodology, a way to combine RPG (Role-Playing-Games) and MABS (Multi-Agent Based Simulation) techniques in an integrated way. The prototype ViP-Jogo Man was designed to be used as a group decision support system in a real problem domain because it encompasses complex negotiation processes and that methodology can be explored to handle conflict resolution.

All sorts of tools are thought to facilitate problem-solving and the engineering of good solutions for the owners of applications. Currently, in informatics at large, complex business tasks are turning into a popular issue, and increasingly there is a need for suitable aids to face several categories of domain-specific interventions. For example, web applications are frequent now in enterprises because they can be accessed via browsers in a standardized way. Also, the development of (mobile) agent-applications for open environments requires heterogeneous platforms to be interoperable, and agents are good for that particular context because they are able to execute and migrate through platforms, interact with each other independently from hosts, and interact with the agents of the application available on different platforms.

# 4     Emerging Trends: From AAMAS02 to AAMAS07

The main trends and drivers are now around the Semantic Web, Web services and service oriented computing, peer-to-peer computing, grid computing, ambient intelligence, self-systems and autonomic computing, and complex systems. There is choreography between technologies and simulation, having the design metaphor at the core of this space of conversations.

According to the AgentLink technology roadmap (2005) the current situation is associated with one design team, agents sharing common goals, closed agent systems applied to specific environments, ad-hoc designs, predefined communication protocols and language, and scalability only allowed in simulation (Luck, 2007). The projections are positive: 35% of software development will use agents and the rate of growth of adoption will increase until 2014 (penetration of 12% by 2010, or a third of long-run adoption level), yet object technologies will be more popular: not all applications of agent technologies will be coined as agent systems. So, in the short term we will have fewer common goals, use of semi-structured agent communication languages (e.g. FIPA ACL), top-down design methodologies (Gaia) compared with today's ad-hoc designs, scalability extended to predetermined and domain-specific environments (today, scalability is only in simulation). In the medium term, we will have the design done by different teams, the use of agreed protocols and languages, standard agent-specific design methodologies, open agent systems in particular domains (bioinformatics, e-commerce), more general scalability, arbitrary numbers and diversity of agents in all such domains, and bridging agents translating between domains. However, we are far from design by diverse teams, truly-open and fully-scalable multi-agent systems, agents capable of learning appropriate communication protocols upon entry to a system, and protocols emerging and evolving through agent interactions.

Looking closer to the World Congresses AAMAS (from 2002 till 2007) we have great difficulties in analyzing the historical evolution of the field and in detecting shifts (new trends), because its organization is based upon possible sessions (space-time availability provided by local organizers) and not on existing tracks (or directions of research). Names attached to sessions are misleading and several papers were included without any reason (more care is required to prepare the sessions, and a new ontology of the field is mostly advisable). This is not the case for the International Joint Conference on AI, where areas are well fixed: from 1981 onwards the main three areas have been knowledge representation, reasoning and machine learning; the other ones go up and down according to their scientific potential and the financial aid assigned, allowing the community to infer the hot topics, shifts and trends.

The agents world is composed of multiple events, mainly workshops, where the community gets more or less strong links. There are also associated regions, like simulation, that adopt a similar form of organization without any well defined geography. For example, the fourth Conference of the European Social Simulation Association (ESSA'07, Toulouse, September 10-14, 2007) was structured into two components (plenary and parallel sessions) and along a continuum (queue) of themes, where social issues (resource sharing, reputation and communication, social influence, feature propagation in a population, market dimensionality, social power structures, historical simulation, rule changes, diffusion of innovation, impact of knowledge on learning, qualitative observation) and applications (policy, firms, organizations, economy, spatial dynamics, opinion and cultural dynamics) are overviewed.

In the mid-90s, there was an attempt made by Yves Demazeau to organize the whole field. He accepted the mainstream ideas of the objects and components movements, and advanced four units: agents (A) to cover the basic elements, environment (E) to cover the passive elements, interactions (I) to cover the means to exchange information and control among agents, and organizations (O) to cover the policies followed to constrain the interactions among agents up to 2003, he developed the idea of a MAS framework and methodology, based upon four vowels A, E, I and O, expanded later on with the U, for the users or the applications. The dynamics chart included the side of decomposition (identification, analysis), of modelling (choice, design) and of tools (programming).

The ATAL (Agents Theory, Architectures and Languages) Workshop was another attempt to characterize the field by dividing it into three main aspects (Wooldridge and Jennings, 1995): the agent contents, its organization and the available tools/infrastructures to build it up.

Another attempt, more general and abstract, to classify the research was made by (Sichman, 2003) along three independent directions (3D scenario): aims (resolution, simulation), description (theories, architectures, languages) and dimensions (agents, environments, interactions and organizations). The goal was to fix exactly each contribution (e.g. models, applications) in the context.

Luck and colleagues (2005), in the AgentLink roadmap, suggested also five components or good tracks of the Agents world: agent level (A), infrastructure and supporting technologies (E), interaction level (I), organization level (O), applications and industry (U), where A, I and O stand for the core set of technologies and techniques required to design and implement agent systems that are now the focus of current research and development.

We took the six conferences of AAMAS (Bologna, Melbourne, New York, Utrecht, Hakodate and Honolulu), from 2002 up to 2007, the whole list of themes (names of the sessions), and adopted the five vowels to analyze the evolution of Agents. Picking up the number of papers (we excluded the posters), attached to each theme, we pictured the state of the art along six consequent years.

The conclusions are the following: low relevance for tools (infrastructures and support technologies) and high relevance for applications (university and industry). The three agent technologies and techniques got 3/4 of the papers with agent or organization level topics at the top. For example, in AAMAS07 the ranking was: applications (25.6%), organization level (25.6%), agent level (23.3%), interaction level (22.6%), and infrastructures and support technologies (3%).

## 5     Future Challenges: Broad and Specific

Future challenges can be classified as broad or specific. Among the broad class, we may find creating tools, techniques and methodologies, automatic specification, development and management, integrating components and features, establishing trade-offs between adaptability and predictability, and linkage with other branches of computer science and other disciplines. Among specific challenges, we may find trust and reputation, virtual organization formation and management, resource allocation and coordination, negotiation, emergence in large-scale agent systems, learning and optimization theory, methodologies, provenance, service architecture and composition, and semantic integration.

Looking to the challenge of trust and reputation we find today: techniques for authentication, verification, validation; sophisticated distributed systems that involve action in absence of strong existing trust relationships; need to address problems of establishing, monitoring and managing trust; interactions in dynamic and open environments; and, trust of agents in agents (norms, reputation, contracts).

Organizing is a complex process, from the bottom-up or top-down point of view, and used in different contexts and applications. Reorganizing is a dynamic process with multiple dimensions and styles that can be predefined using an organizational language. Multi-agent organizations for current and next applications need to combine agent centered and organization points of view, to combine agent level and system level programming of organizational models, and to integrate and enforce dynamic and adaptive organizations.

An interesting direction is to design complex systems that evolve in changing environments with MAS. Self-organization is a promising paradigm to build up these systems adaptive where the collective function arises from the local interactions and the whole design grows bottom-up. The difficulty appears when discovering the right behaviours at the agent-level to make the global function emerge. Often, simulation helps designers to trap the correct behaviours during the design stage.

A complex problem was one of the most frequent targets chosen by the AAMAS-07 authors, potential candidates for the best paper. This is not strange because nowadays agent technologies are selected by large enterprises to help the problem-solving, yet building such a software system to achieve complex tasks is a rather difficult job because classical engineering techniques are not always useful. One of the hardest cases happens when the system must be composed of a large number of interacting entities and the global behaviour is said to emerge. The difficulty consists in linking the goal defined at the macroscopic level with the corresponding behaviour at the microscopic level.

Air traffic flow management is one of the fundamental challenges facing the Federal Aviation Administration (FAA) today. Finding reliable and adaptive solutions to the flow management problem is of paramount importance if the Next Generation Air Transportation Systems are to achieve the stated goal of accommodating three times the current traffic volume. This problem (Tumer and Agogino, 2007) is particularly complex as it requires the integration and/or coordination of many factors including: new data (e.g., changing weather info), potentially conflicting priorities (e.g., different airlines), limited resources (e.g., air traffic controllers) and very heavy traffic volume (e.g., over 40,000 flights over the US airspace). The air traffic flow simulator developed at NASA, used extensively by the FAA and industry, was selected to test a multi-agent algorithm for traffic flow management. An agent was associated with a fix (a specific location in 2D space) and its action consisted of setting the separation required among the airplanes going through that fix. Agents used reinforcement learning to set this separation and their actions speed up or slow down traffic to manage congestion. The results showed that agents receiving personalized rewards reduced congestion by up to 45% over agents receiving a global reward and by up to 67% over a current industry approach (Monte Carlo estimation).

Argumentation in agent systems has several hot research issues now, for example the study of protocol properties, combination of dialogue types (deliberation), multi-party dialogues, protocol design articulated with agent design, embedding in social context, and a framework for dialogue games.

Communication will open new avenues for research along the following directions: concepts for the semantics of both messages and conversations; standard languages for protocol description; tools for supporting a conversation policy construction and implementation; a complete semantics for agent communication; and, focus on multi-party communication.

Trust and reputation mechanisms are used to infer expectations of future behaviour from past interactions (Hermoso et al., 2006), for example in peer-to-peer systems. They are of particular relevance when agents have to choose appropriate counterparts for their interactions (e.g. reputation values about third parties), as it occurs with virtual (and regulated) organizations. The use of organizational structures helps to solve this difficulty and improves the efficiency of trust and reputation mechanisms by endowing agents with extra data to select the best agents to interact with. Certain structural properties of virtual organizations (e.g. governor agents, filters, protocols of sequential actions) able to limit the freedom of choice of agents are used to build an efficient trust model in a local way.

New applications need more research on mastering decentralized coordination and autonomic, adaptable systems, supporting heterogeneity, which involves dealing with semantics as a prerequisite for the quantum leap and based on infrastructure standards, modeling systems with parallel and possibly inconsistent goals and tasks, and exploration of Web 2.0 (e.g. $2^{nd}$ Life) as a playground for agents. The bet on ambient intelligence and Internet will be, for sure, a driver of new visions (decentralized setting, many devices, many users, dynamic sensor-actor networks, and a huge need for interaction).

Finally, there is a theme, serious games (SGs), that deserves to be put into focus, because it opens an interesting range of useful applications (e.g. e-democracy for integrated coastal area management, to support electronic debates and the implementation of contracts). SGs are a type of software developed with game technology and game design principles, for a primary purpose other than pure entertainment. Creating games that simulate functional entities (e.g. battles, but also processes and events at large) can be interesting not only for education (enterprise learning, professional training), but also for group-decision management (business intelligence), advertising, healthcare (see PortEdu portal, AMPLIA learning environment), public policy generation, or even politics. How can institutional or social power be structured, as part of the traditional hierarchies? Can organizational-dynamic games be of any utility for understanding democracy, the role of individuals and collectives and their relations with power? Machiavelli, Spinoza and Hobbes thought multitudes (a cooperative convergence of subjects) could be a substitute for groups dominated by a sole leader and support for constituent power. Subjects articulated via a network may generate a productive force and advance new ideas by working together, without any central boss. So, a whole of singularity becomes more productive (innovative) than the mass of individuals (representing a false unity).

The multitude or singular multiplicity presents advantages over other types of organization because each subject can speak for himself and cooperate with the others. As a matter of fact, the multitude is an active social agent, a multiplicity that acts (and works, disobeys) and a form of composition or contract among active individuals.

# 6    Conclusions

Distributed environments have grown dramatically creating specializations (e.g. various facets of multi-agent systems, semantic web communities) and technical (and technological) demands. It is important to identify key tracks and spaces of intersection between these expanding research areas, in order to discover synergies (points of confluence), redundancies, or difficult problems. If there is really a field of agents (at large), it will be desirable to draw a map containing its geographical elements (the so-called new set of categories to assemble the large queue of themes). So, it would be easier to organize new programmes of research, where strands of inquiry could be woven together to attack new problems. From example, the area of peer-to-peer knowledge sharing has a technological demand for more controllable, accessible, faster and larger scale systems. But this presents a challenge to traditional multi-agent coordination, ontology management and service architectures.

A nice example of a distributed environment is the case of autonomic electronic institutions which proved to be valuable to regulate open agent systems or agent societies. The rules of the game are fixed, i.e. it is known what agents are permitted and forbidden to do and under what circumstances, and the institutions can adapt their regulations to comply with their goals despite coping with varying populations of self-interested external agents. Self-management (organization, configuration, diagnosis, repair) becomes an interesting feature to face varying agents' behaviours and helpful to support simulated agent populations.

The geography proposed by the AgentLink roadmap of (Luck et al., 2005) is so far the fastest way to analyze the evolution of the field of autonomous agents, and we applied it to the last two editions of AAMAS Conferences, where we found a stable situation concerning the main tracks of organization and agent levels.

There are increasing analogies between open problems in pairs of informatics areas, such as distributed systems and agent-based social simulation. Questions regarding micro to macro, cooperation and trust, or evolving network structures are central to software engineering, and modeling methodologies may be combined productively. Reputation, developed within electronic auctions, is also a key tool for partner selection, and can easily be exported into other domains from social networks to institutional evaluation. There is also interweaving of human and technological communities (e.g. ambient intelligence) during the construction of connected communities where software entities act on behalf of users and cooperate with infohabitants. These intersections occur often in the organization of conference sessions, where a paper could be classified under different categories and was inserted in a session on account of room availability. Such phenomena disturb the rankings and may falsify the conclusions.

**Acknowledgements.** A period of study at the Politechnic School of the University of São Paulo (USP) and the interaction with my colleague Jaime Sichman allowed the writing of this chapter. This work was financially supported by the LabMAg R&D unit, under the Portuguese FCT, and by the Brazilian CNPq BEV n° 170104/07-9. The chapter does not necessarily reflect the views of these two agencies.

# References

Adamatti, D.F., Sichman, J.S., Coelho, H.: Using Virtual Players in GMABS Methodology: A Case Study in Natural Resources Management. In: Proceedings of the 4th Conference of the European Social Simulation Association (ESSA'07), Toulouse, September 10-14 (2007)

Agotnes, T., Walther, D.: Towards a Logic of Strategic Ability under Bounded Memory. In: Proceedings of the Multi-Agent Logics, Languages, and Organisations Federated Workshops (MALLOW'007), Languages, methodologies and Development tools for multi-agent systems (LADS'007) Workshop, Durham, September 3-7 (2007)

Alechina, N., Logan, B.: A Logic of Situated Resource-Bounded Agents. In: Proceedings of the Multi-Agent Logics, Languages, and Organisations Federated Workshops (MALLOW'007), Languages, methodologies and Development tools for multi-agent systems (LADS'007) Workshop, Durham, September 3-7 (2007)

Badr, I., Mubarak, H., Göhner, P.: Extending the MaSE Methodology for the Development of Embedded Real-Time Systems. In: Proceedings of the Multi-Agent Logics, Languages, and Organisations Federated Workshops (MALLOW'007), Languages, methodologies and Development tools for multi-agent systems (LADS'007) Workshop, Durham, September 3-7 (2007)

Costa, A.R., Dimuro, G.P.: Semantical Concepts for a Formal Structural Dynamics of Situated Multiagent Systems. In: Proceedings of the Multi-Agent Logics, Languages, and Organisations Federated Workshops (MALLOW'007), Coordination, Organization, Institutions and Norms in Agents Systems (COIN'007) Workshop, Durham, September 3-7 (2007)

Demazeau, Y.: Multi-Agent Systems Methodology. CNRS Leibniz IMAG, Rennes (Oct. 2003)

Dignum, V., Dignum, F.: A Logic for Agent Organizations. In: Proceedings of the Multi-Agent Logics, Languages, and Organisations Federated Workshops (MALLOW'007), Formal Approaches to Multi-Agent Systems (FAMAS'007) Workshop, Durham, September 3-7 (2007)

Georgeff, M., Ingrand, F.F.: Real-time reasoning: the monitoring and control of spacecraft systems. In: Sixth Conference on Artificial Intelligence Applications, May 5-9 (1990)

Hepple, A., Dennis, L., Fisher, M.: A Common Basis for Agent Organisation BDI Languages. In: Proceedings of the Multi-Agent Logics, Languages, and Organisations Federated Workshops (MALLOW'007), Languages, methodologies and Development tools for multi-agent systemS (LADS'007) Workshop, Durham, September 3-7 (2007)

Hermoso, R., Billhardt, H., Centeno, R., Ossowski, S.: Effective use of organisational abstractions for confidence models. In: O'Hare, G.M.P., Ricci, A., O'Grady, M.J., Dikenelli, O. (eds.) ESAW 2006. LNCS (LNAI), vol. 4457, pp. 368-383. Springer, Heidelberg (2007)

Klügl, F.: Measuring Complexity of Multi-Agent Simulations, an Attempt using Metrics. In: Proceedings of the Multi-Agent Logics, Languages, and Organisations Federated Workshops (MALLOW'007), Languages, methodologies and Development tools for multi-agent systemS (LADS'007) Workshop, Durham, September 3-7 (2007)

Lloyd, J.W., Ng, K.S.: Probabilistic and Logical Beliefs. In: Proceedings of the Multi-Agent Logics, Languages, and Organisations Federated Workshops (MALLOW'007), Languages, methodologies and Development tools for multi-agent systems (LADS'007) Workshop, Durham, September 3-7 (2007)

Lorini, E., Herzig, A., Broersen, J., Troquard, N.: Grounding Power on Actions and Mental Attitudes. In: Proceedings of the Multi-Agent Logics, Languages, and Organisations Federated Workshops (MALLOW'007), Formal Approaches to Multi-Agent Systems (FAMAS'007) Workshop, Durham, September 3-7 (2007)

Luck, M., McBurney, P., Shehory, O., Willmott, S., et al.: Agent Technology: Computing as Interaction, A Roadmap for Agent Based Computing, AgentLink III (2005)

Luck, M.: General Introduction to Multi-Agent Systems. Tutorial slides of EASSS2007, Ninth edition of the European Agent Systems Summer School (EASSS), Durham, August 27-31 (2007)

Nakayama, L., Vicari, R.M., Coelho, H.: An Information Retrieving Service for Distance Learning. The IPSI BgDTransactions on Internet Research 1(1) (2005)

Sichman, J.S.: Raciocínio Social e Organizacional em Sistemas Multiagentes: Avanços e Perspectivas. Tese de Livre-Docência, Escola Politécnica da USP, Brasil (2003)

Silva, P.T., Coelho, H.: An Hybrid Approach to Teamwork. In: Proceedings of VI Encontro Nacional de Inteligência Artificial (ENIA2007), XXVII Congresso da SBC, Rio de Janeiro, July 2-6 (2007)

Suárez-Romero, J., Alonso-Betanzos, A., Guijarro-Berdiñas, B.: Integrating a Priority-Based Scheduler of Behaviours in JADE. In: O'Hare, G., Ricci, A., O'Grady, M., Dikenelli, O. (eds.) Proceedings of the 7th International Workshop on Engineering Societies in the Agents World (ESAW06), Dublin, September 6-8 (2006)

Tumer, K., Agogino, K.: Distributed Agent-Based Air Traffic Flow Management. In: Proceedings of the Sixth International Joint Conference on Autonomous Agents and Multi-Agent Systems (AAMAS-07), Honolulu, July 16-18 (2007)

Wegner, P.: Models and Paradigms of Interaction. OOPSLA Tutorial Notes (1995)

Wegner, P.: Why Interaction is More Powerful than Algorithms. Communications of the ACM 40(5), 80–91 (1997)

Wooldridge, M., Jennings, N.: Intelligent Agents: Theory and Practice. The Knowledge Engineering Review 10(2) (1995)

# Affective Intelligence: The Human Face of AI

Lori Malatesta, Kostas Karpouzis, and Amaryllis Raouzaiou

National Technical University of Athens 15780, Zographou, Athens, Greece
{kkarpou, lori, araouz}@image.ntua.gr

**Abstract.** Affective computing has been an extremely active research and development area for some years now, with some of the early results already starting to be integrated in human-computer interaction systems. Driven mainly by research initiatives in Europe, USA and Japan and accelerated by the abundance of processing power and low-cost, unintrusive sensors like cameras and microphones, affective computing functions in an interdisciplinary fashion, sharing concepts from diverse fields, such as signal processing and computer vision, psychology and behavioral sciences, human-computer interaction and design, machine learning, and so on. In order to form relations between low-level input signals and features to high-level concepts such as emotions or moods, one needs to take into account the multitude of psychology and representation theories and research findings related to them and deploy machine learning techniques to actually form computational models of those. This chapter elaborates on the concepts related to affective computing, how these can be connected to measurable features via representation models and how they can be integrated into human-centric applications.

## 1 Introduction

As research has revealed the deep role that emotion and emotional expression play in human social interaction, researchers in human computer interaction have proposed that more effective human computer interfaces can be realized if the interface models the user's emotion as well as expresses emotions. Affective computing is computing that relates to, arises from, or deliberately influences emotion or other affective phenomena. According to Rosalind Picard's pioneering article [32], if we want computers to be genuinely intelligent and to interact naturally with us, we must give computers the ability to recognize, understand, and even to have and express emotions. These positions have become the foundations of research in the area and have been investigated in great depth after their first postulation.

Emotion is fundamental to human experience, influencing cognition, perception, and everyday tasks such as learning, communication, and even rational decision-making. Affective computing aspires to bridge the gap that typical human

M. Bramer (Ed.): Artificial Intelligence, LNAI 5640, pp. 53–70, 2009.

computer interaction largely ignored thus creating an often frustrating experience for people, in part because affect had been overlooked or was hard to measure.

In order to take these ideas a step further, towards the objectives of practical applications, we need to adapt methods of modelling affect to the requirements of the project's showcases. To do so it is fundamental to review prevalent psychology theories on emotion, to disambiguate their terminology and identify the fitting computational models that can allow for affective interactions in the desired environments.

## 2     Terminology Disambiguation

We speak of disambiguation since a lot of confusion exists regarding emotion research terminology, and not without a reason. Different definitions of the role and nature of emotions arise from different scientific approaches since emotion research is typically multidisciplinary. Different disciplines (i.e. psychology, cognitive neuroscience etc) provide theories and corresponding models that are based on diverse underlying assumptions, are based on different levels of abstraction and may even have different research goals altogether.

So what are emotions? It largely remains an open question. Some define it as the physiological changes caused in our body, while others treat it as a purely intellectual thought-process.

In psychology research [36] the term *'affect'* is very broad, and has been used to cover a wide variety of experiences such as emotions, moods, and preferences. In contrast, the term *'emotion'* [9] tends to be used to refer to fairly brief but intense experiences although it is also used in a broader sense. Finally, moods or states describe low-intensity but more prolonged experiences.

From a cognitive neuroscience point of view, Damasio [10] makes a distinction between emotions, which are publicly observable body states, and feelings, which are mental events observable only to the person having them. Based on neuroscience research he and others have done, Damasio argues that an episode of emotion begins with an emotionally "competent" stimulus (such as an attractive person or a scary house) that the organism automatically appraises as conducive to survival or well-being (a good thing) or not conducive (bad). This appraisal takes the form of a complex array of physiological reactions (e.g., quickening heartbeat, tensing facial muscles), which is mapped in the brain. From that map, a feeling arises as "an idea of the body when it is perturbed by the emoting process" [10].

It is apparent that there is no right or wrong approach, and an attempt at a full terminology disambiguation would not be possible without biasing our choices towards one theory over the other. This is to make the point that the context of each approach has to be carefully defined. Next we are going to enumerate core elements of emotion and ways to distinguish them from other affective phenomena. This will lead us to a short description of the directions of affective computing. Subsequently we will put forward the most prevalent psychological theories of emotion along with corresponding computational modelling approaches and couple them to the affective computing goals and more specifically to the goals of practical applications.

## 2.1   Defining 'Emotion' and 'Feeling'

Emotion, according to Klaus Scherer ([40], [42]), can be defined as an episode of interrelated, synchronized changes in the states of all or most of five organismic subsystems in response to the evaluation of an external or internal stimulus event as relevant to major concerns of the organism. The components of an emotion episode are the particular states of the subsystems mentioned. The process consists of the coordinated changes over time.

Most current psychological theories postulate that subjective experience, peripheral physiological response patterns, and motor expression are major components of emotion. These three components have often been called the emotional response triad. Some theorists include the cognitive and motivational domains as components of the emotion process. The elicitation of action tendencies and the preparation of action have also been implicitly associated with emotional arousal. However, only after explicit inclusion of motivational consequences in theories (and Frijda's forceful claim for the emotion-differentiating function of action tendencies, see [14]), have these important features of emotion acquired the status of a major component. The inclusion of a cognitive information-processing component has met with less consensus. Many theorists still prefer to see emotion and cognition as two independent but interacting systems. However, one can argue that all subsystems underlying emotion components function independently much of the time, and that the special nature of emotion as a hypothetical construct consists of the coordination and synchronization of all these systems during an emotion episode [43].

How can emotions, as defined above, be distinguished from other affective phenomena such as feelings, moods, or attitudes? Let us take the term feeling first. Scherer aligns feeling with the "subjective emotional experience" component of emotion, thus reflecting the total pattern of cognitive appraisal as well as motivational and somatic response patterning that underlie the subjective experience of an emotion. If we use the term feeling, a single component denoting subjective experience process, as a synonym for emotion (the total multi-modal component process), this is likely to produce serious confusion and hamper our understanding of the phenomenon.

If we accept feeling as one of emotion's components, then the next step is to differentiate emotion from other types of affective phenomena. Instances of these phenomena, which can vary in degree of affectivity, are often called "emotions" in the literature. According to Scherer [44], there are five such types of affective phenomena that should be distinguished from emotion: *preferences*, *attitudes*, *moods*, *affective dispositions* and *interpersonal stances*.

In order to differentiate emotions from the rest of the affective phenomena we shall sketch out core elements of emotions.

### 2.1.1  Event Focus

Emotions are generally elicited by stimulus events. Something happens to the organism, which, after having been evaluated for its significance, stimulates or triggers a response. Often such events will consist of natural phenomena like thunderstorms, or the behaviour of other people or animals that may have significance for our well-being. In other cases, one's own behaviour can be the event that elicits emotion, as in the case of pride, guilt, or shame. In addition to such events that are more or less external to the organism, internal events are explicitly considered as emotion elicitors. These may consist of sudden neuro-endocrine or physiological changes or, more typically, of memories or images that might come to our mind. These recalled or imagined representations of events can be sufficient to generate strong emotions [18]. The event focus element means that emotions need to be somehow connected or anchored to a *specific event*, external or internal, rather than being free-floating, resulting from a strategic or intentional decision, or existing as a permanent characteristic of an individual.

### 2.1.2  Appraisal Basis

A central aspect of the component process definition of emotion is that the eliciting event and its consequences must be relevant to major concerns of the individual. This seems rather obvious; as we do not generally get emotional about things or people we do not care about. Frijda [14] talks of emotions as relevance detectors. Componential theories of emotion generally assume that the relevance of an event is determined by a rather complex, yet very rapidly occurring evaluation process that can take place on several levels of processing, ranging from automatic and implicit to conscious conceptual or propositional evaluations [27]. It makes sense to distinguish between intrinsic and extrinsic appraisal. Intrinsic appraisal evaluates the features of an object or person independently of the current needs and goals of the appraiser, based on genetic preferences (e.g. sweet taste) or learned preferences (e.g., bittersweet food) (see [40], [41]). Extrinsic appraisal (also known as transactional appraisal; see [26]) evaluates events and their consequences in terms of their contribution to the salient needs, desires, or goals of the appraiser.

### 2.1.3  Response Synchronization

This element is also implied by the adaptive functions of emotion. If emotions prepare appropriate responses to events, the response patterns must correspond to the appraisal analysis of the presumed implications of the event. Given the importance of the eliciting event, which disrupts the flow of behaviour, all or most of the subsystems of the organism must contribute to response preparation. The resulting massive mobilization of resources must be coordinated, a process which can be described as response synchronization [42]. This is in fact one of the most important design features of emotion, one that in principle can be operationalized and measured empirically.

### 2.1.4  Rapidity of Change

Events, and particularly their appraisal, change rapidly, often because of new information, or due to re-evaluation. As appraisal drives the patterning of the responses in the interest of adaptation, the emotional response patterning is also likely to change rapidly. While we are in the habit of talking about "emotional states", these are rarely steady states. Rather, emotion processes are undergoing constant modification, allowing rapid readjustment to changing circumstances or evaluations.

### 2.1.5  Behavioural Impact

Emotions prepare the ground for adaptive action tendencies and their motivational underpinnings. In this sense, they have a strong effect on behaviour resulting from emotion. They often interrupt ongoing behaviour sequences, and generate new goals and plans. In addition, the motor expression component of emotion has a strong impact on communication. This may have important consequences for social interaction.

### 2.1.6  Intensity

Given the importance of emotions for behavioural adaptation, one can assume the intensity of the response patterns and the corresponding emotional experience to be relatively high. This may be an important design feature in distinguishing emotions from moods, for example.

### 2.1.7  Duration

As emotions imply massive response mobilization and synchronization as part of specific action tendencies, their duration must be relatively short in order not to tax the resources of the organism, and to allow behavioural flexibility. In contrast, low-intensity moods that have little impact on behaviour can be maintained for much longer periods without there being adverse effects.

## 3     Distinguishing 'Emotion' from Other Affective Phenomena

### 3.1     Features of Other Affective Phenomena

Having presented the basic elements of emotions it is now possible to define the other phenomena mentioned earlier in such a way as to distinguish them from emotions.

### 3.1.1  Preferences

We will refer to relatively stable evaluative judgments in the sense of liking or disliking a stimulus, or preferring it to other objects or stimuli, as *preferences*. By definition, stable preferences should generate intrinsic appraisal independently of current needs or goals, although the latter might modulate the appraisal [41]. The affective states produced by encountering attractive or aversive stimuli (event focus) are stable, of relatively low intensity, and do not produce pronounced response synchronization. Preferences generate unspecific positive or negative feelings, with low behavioural impact, except for tendencies towards approach or avoidance.

### 3.1.2  Attitudes

Relatively enduring beliefs and predispositions towards specific objects or persons are generally called attitudes. Social psychologists have long identified three components of attitudes (see [4]): a) a cognitive component (beliefs about the attitude object), b) an affective component (consisting mostly of differential valence), and c) a motivational or behavioural component (a stable action tendency with respect to the object, e.g., approach or avoidance). Attitude objects can be things, events, persons, and groups or categories of individuals. Attitudes do not need to be triggered by event appraisals, although they may become more salient when encountering or thinking of the attitude object. The affective states induced by a salient attitude can be labelled by terms such as "hating", "valuing", or "desiring". Intensity and response synchronization are generally weak, and behavioural tendencies are often overridden by situational constraints.

### 3.1.3  Moods

Emotion psychologists have often discussed the difference between mood and emotion (e.g. [15]). Generally, moods are considered as diffuse affect states, characterized by a relatively enduring predominance of certain types of subjective feelings that affect the experience and behaviour of a person. Moods may often develop without an apparent cause that could be clearly linked to an event or specific appraisal. They are generally of low intensity and show little response synchronization, but may last for hours or even days. Examples are being cheerful, gloomy, listless, depressed, or buoyant.

### 3.1.4  Affective Dispositions

Many stable personality traits and behaviour tendencies have a strong affective core (e.g., being nervous, anxious, irritable, reckless, morose, hostile, envious or jealous). These dispositions describe the tendency of a person to experience certain moods more frequently or to be prone to react with certain types of emotions, even upon slight provocation. Not surprisingly, terms like "irritable" or "anxious" can describe both affect dispositions and momentary moods or emotions. It is important to specify whether the term is used to qualify personality disposition or an episodic state. Affect dispositions also include emotional pathology: while being in a depressed mood is quite normal, being constantly depressed may be a sign of an affective disturbance requiring medical attention.

### 3.1.5  Interpersonal Stances

This category refers to an affective style that spontaneously develops or is strategically employed in the interaction with a person or a group of persons, colouring the interpersonal exchange in that situation (e.g. being polite, distant, cold, warm, supportive, contemptuous). Interpersonal stances are often triggered by events, such as encountering a certain person. However, they are less shaped by spontaneous appraisal than by affect dispositions, interpersonal attitudes, and, most importantly, strategic intentions. Thus, when an irritable person encounters a disliked individual, that person is more likely to adopt an interpersonal stance of hostility in the interaction, as compared to an agreeable person.

### 3.2    Emotions in Applied Intelligence

Having distinguished emotions from other types of affective phenomena it is now of particular interest, in regard to the new media domain, to present a suggested distinction on a different level. Scherer [43] questioned the need to distinguish between two different types of emotion: (1) *aesthetic emotions* (2) *utilitarian emotions*. The latter correspond to the "garden variety" of emotions usually studied in emotion research, such as anger, fear, joy, disgust, sadness, shame, guilt. These types of emotions can be considered utilitarian in the sense of facilitating our adaptation to events that have important consequences for our well-being. Such adaptive functions are the preparation of action tendencies (fight, flight), recovery and reorientation (grief, work), motivational enhancement (joy, pride), or the creation of social obligations (reparation). Because of their importance for survival and well-being, many utilitarian emotions are high-intensity emergency reactions, involving the synchronization of many subsystems, as described earlier. In the case of aesthetic emotions, adaptation to an event that requires the appraisal of goal relevance and coping potential is absent, or much less pronounced. Kant defined aesthetic experience as "disinterested pleasure" ([24]), highlighting the complete absence of utilitarian considerations. Thus, my aesthetic experience of a work of art or a piece of music is not shaped by the appraisal of the work's ability to satisfy my bodily needs, further my current goals or plans, or correspond to my social values. Rather, aesthetic emotions are produced by the appreciation of the intrinsic qualities of a work of art or an artistic performance, or the beauty of nature. Examples of such aesthetic emotions are: being moved or awed, full of wonder, admiration, bliss, ecstasy, fascination, harmony, rapture, solemnity.

This differentiation of emotions has an impact on the way an appraisal based modelling approach would be implemented. It would not make sense to try and model all the proposed components of an appraisal process in cases where only aesthetic emotions are expected. On the other hand, it would make sense to provide a deeper model in cases where anger or frustration are common emotional states such as in the example of interactive Television.

## 4    Areas of Affective Computing

Affective computing deals with the design of systems and devices which can recognize, interpret, and process emotions. We are going to fledge out the potentials this research domain can provide in the field of new media applications and identify the matching theoretical background that will act as a tool for effectively modelling emotional interaction in such environments.

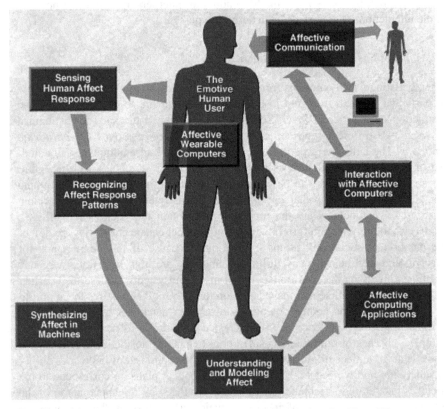

**Fig. 1.** The research areas of affective computing as visualized by MIT

## 4.1    Detecting and Recognizing Emotional Information

Detecting emotional information usually involves passive sensors which capture data about the user's physical state or behaviour. The data gathered is often analogous to the cues humans use to perceive emotions in others. For example, a video camera might capture facial expressions, body posture and gestures, while a microphone might capture speech. Other sensors detect emotional cues by directly measuring physiological data, such as skin temperature and galvanic resistance.

Recognizing emotional information requires the extraction of meaningful patterns from the gathered data ([21], [23]). This is done by parsing the data through various processes such as facial expression detection, gesture recognition, speech recognition, or natural language processing.

## 4.2    Emotion in Machines

By emotion in machines, we refer to the simulation of emotions. The goal of such simulation is to enrich and facilitate interactivity between human and machine. The most common and probably most complicated application of this simulation lies in the field of conversational agents. Such a simulation is closely coupled with emotional understanding and modelling as explained below. This being said it is important to mention that less sophisticated simulation approaches often produce surprisingly engaging experiences in the area of new media. It is often the case that our aim is not to fully simulate human behaviour and emotional responses, but merely to illustrate emotion in a pseudo-intelligent way that makes sense in the specific context of interaction.

## 4.3    Emotional Understanding

Emotional understanding refers to the ability of a device to not only detect emotional or affective information, but also to store, process, build and maintain an emotional model of the user. The goal is to understand contextual information about the user and her environment, and formulate an appropriate response. This is difficult because human emotions arise from complex external and internal contexts [17].

Possible features of a system which displays emotional understanding might be adaptive behaviour, for example, avoiding interaction with a user it perceives to be angry. In the case of affect-aware applications, emotional understanding makes sense in tracking the user's emotional state and adapting environment variables according to the state recognised. Questions regarding the level of detail of the tracking performed, the theoretical grounds for the analysis of the data collected and the types of potential output that would make sense for such an interactive process, are paramount.

## 5     Emotion Descriptions and Emotion Models

Having reviewed the areas of affective computing, it is time to start focusing on the available theories, descriptions and models that can support these goals. We start with reviewing the three big groups of emotion descriptions as identified by the members of the Humaine (Human-Machine Interaction Network on Emotion) Network of Excellence.

It is important to stress the difference that exists between emotion models and emotion descriptions. By emotion descriptions we refer to different ways of representing emotions and their underlying psychological theories whereas with the term emotional models we talk about the computational modelling of these theories in specific context.

### 5.1     Categorical Representations

Categorical representations are the simplest and most widespread, using a word to describe an emotional state. Such category sets have been proposed on different grounds, including evolutionarily basic emotion categories; most frequent everyday emotions; application-specific emotion sets; or categories describing other affective states, such as moods or interpersonal stances. (Feeltrace core vocabulary in [8]; Ortony's list of emotion words in [30] and [31]; Ekman's list of six basic emotions in [12])

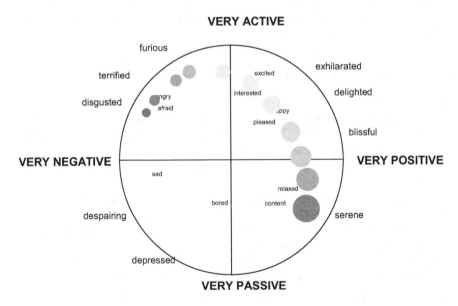

**Fig. 2.** A dimensional representation of emotion

## 5.2   Dimensional Descriptions

Dimensional descriptions capture essential properties of emotional states, such as arousal (active/passive) and valence (negative/positive). Emotion dimensions can be used to describe general emotional tendencies, including low-intensity emotions.

In addition to these two, there are a number of other possible dimensions, such as power, control, or approach / avoidance, which add some refinement. The most obvious is the ability to distinguish between fear and anger, both of which involve negative valence and high activation. In anger, the subject of the emotion feels that he or she is in control; in fear, control is felt to lie elsewhere.

Dimensional representations are attractive mainly because they provide a way of describing emotional states that is more tractable than using words. This is of particular importance when dealing with naturalistic data, where a wide range of emotional states occur. Similarly, they are much more able to deal with non-discrete emotions and variations in emotional state over time. A further attraction is the fact that dimensional descriptions can be translated into and out of verbal descriptions. This is possible because emotion words can, to an extent, be understood as referring to positions in activation-evaluation space.

## 5.3   Appraisal Theories and Representations

Appraisal theories focus on the emotion elicitation process in contrast with the previously mentioned approaches that emphasize the consequences/ symptoms of an emotional episode. Appraisal representations characterise emotional states in terms of the detailed evaluations of eliciting conditions, such as their familiarity, intrinsic pleasantness, or relevance to one's goals. Such detail can be used to characterise the cause or object of an emotion as it arises from the context, or to predict emotions in AI systems ([26],[14], [40]).

Appraisal theories are very common in emotion modelling since their structure makes it feasible for simulating their postulations in computational models. Moreover, it is often the case that an appraisal theory was formulated explicitly in order to be implemented in a computer. Such an example is the OCC theory ([30]). This is sometimes a source of confusion, since the underlying emotion theory is unavoidably very closely linked with the actual modelling approach.

According to cognitive theories of emotion ([25]), emotions are closely related to the situation that is being experienced (or, indeed, imagined) by the agent. In the following, we are going to outline four of the most prevalent theories in the field.

### 5.3.1   OCC Theory

The theory of Ortony, Clore and Collins ([30]) assumes that emotions develop as a consequence of certain cognitions and interpretations. Therefore, it exclusively concentrates on the cognitive elicitors of emotions. The authors postulate that three aspects determine these cognitions: events, agents, and objects. Emotions represent valenced (positive/ negative) reactions to these perceptions of the world. One can be pleased about the consequences of an event or not (pleased/displeased); one can endorse or reject the actions of an agent (approve/disapprove) or one can like or not like aspects of an object (like/dislike).

A further differentiation consists of the fact that events can have consequences for others or for oneself and that an acting agent can be another or oneself. The consequences of an event for another can be divided into desirable and undesirable; the consequences for oneself as relevant or irrelevant expectations. Relevant expectations for oneself finally can be differentiated again according to whether they actually occur or not (confirmed/disconfirmed).

With the help of such a formal system, a computer should be able to draw conclusions about emotional episodes which are presented to it. The authors are not interested in the question if machines are actually 'experiencing' emotions. Rather, they only focus on the ability to understand emotions, reason about them and express them.

### 5.3.2   Scherer's Appraisal Theory

Scherer's appraisal theory ([40], [42]) is more commonly known as the component process model. For Scherer five functionally defined subsystems are involved with emotional processes. An information-processing subsystem evaluates the stimulus through perception, memory, forecast and evaluation of available information. A supporting subsystem adjusts the internal condition through control of neuroendocrine, somatic and autonomous states. A leading subsystem plans, prepares actions and selects amongst competitive motives. An acting subsystem controls motor expression and visible behaviour. A monitor subsystem finally controls the attention assigned to the present states and passes the resulting feedback on to the other subsystems.

Scherer is especially interested in the information-processing subsystem. According to his theory, this subsystem is based on appraisals which Scherer calls stimulus evaluation checks (SEC). The result of these SECs causes again changes in the other subsystems.

Scherer enumerates five substantial SECs, four of which possess further sub-checks. The novelty check decides whether external or internal stimuli have changed; its sub-checks are suddenness, confidence and predictability. The intrinsic pleasantness check specifies whether the attraction is pleasant or unpleasant and causes appropriate approximation or avoidance tendencies. The goal significance check decides whether the event supports or prevents the goals of the person; its sub-checks are goal relevance, probability of result, expectation, support character and urgency. The coping potential check determines to what extent the person believes he or she has events under control; its sub-checks are agent, motive, control, power and adaptability. The compatibility check finally compares the event with internal and external standards; its sub-checks are externality and internality.

According to Scherer, each emotion can thus be clearly determined by a combination of the SECs and sub-checks. An appropriate table with such allocations can be found in ([41]).

### 5.3.3  Roseman's Theory

Roseman first presented his theory in the late seventies and then modified it several times in later years. Five cognitive dimensions determine whether an emotion arises and which one it is. These dimensions were re-visited several times. Nevertheless the model proposed was never empirically validated thus leading to new revisions/ additions ([34], [35]).

### 5.3.4  Frijda's Theory

Frijda [14, 15] points out that the word "emotion" does not refer to a "natural class" and that it is not able to refer to a well-defined class of phenomena which are clearly distinguishable from other mental and behaviour events. For him, therefore, the process of emotion emergence is of larger interest.

The centre of Frijda's theory is the term concern. A concern is the disposition of a system to prefer certain states of the environment and of its own organism over the absence of such conditions. Concerns produce goals and preferences for a system. If the system has problems in realizing these concerns, emotions develop.

The strength of such an emotion is determined essentially by the strength of the relevant concern(s). For Frijda, emotions are necessary for systems that realize multiple concerns in an uncertain environment. If a situation occurs, in which the realization of these concerns appears endangered, so-called action tendencies develop. These action tendencies are linked closely with emotional states and serve as a safety device for what Frijda calls concern realization (CR).

## 6    Computational Models

Emotion models are computational approaches that are based on these descriptions/ theories (often combinations of more than one type of description). They aspire to validate them and possibly to extend them. They allow the simulation of behaviour and aid in both recognising and understanding human emotions as well as generating synthetic emotional responses. Keeping in mind these descriptions, emotion models can be divided into two categories:

- ones that take into account the situations that initiate the emotions and how they are construed by the experiencer and focus on the predicted emotion – from now on we shall refer to them as '*deep models*'.

- ones that deal with the 'results' of an emotional episode i.e. facial expression/ voice etc. – from now on we shall refer to them as '*shallow models*'.

What is the motivation for the development of a computational model implementing a particular emotion theory, or attempting to account for particular data? On a research level, it provides an opportunity for validation of the theory's claims. It also allows for the generation of alternative hypotheses explaining specific affective data or phenomena. The aim of this chapter is to investigate existing modelling approaches and to isolate the ones that would potentially meet a particular set of requirements. In order to do that we shall first go through a general overview of emotion modelling literature.

A number of computational models addressing emotion have been developed in cognitive science and AI. These models range from individual processes to integrated architectures, and explore several of the emotion theories outlined above. One thing that differentiates these modelling approaches is the *level of abstraction*. At the higher level of abstraction are architecture-level models that embody emotional processing. At an intermediate level of abstraction are task-level models of emotion, which focus on addressing a single task, such as natural language understanding or specific problem solving. At lower levels of abstraction are mechanism-level models, which attempt to emulate some specific aspect of affective processing. The level of abstraction is found to be a key criterion in the selection of the appropriate models for an actual application.

According to a review on emotional models by Hudlicka ([19]) the most frequently modelled process has been cognitive appraisal, whereby external and internal stimuli (emotion elicitors) are mapped onto a particular emotion. Several alternatives have been hypothesized for these processes in the psychological literature ([14], [26], [43], [30], [39]). A number of these models have been implemented, both as stand-alone versions, and integrated within larger agent architectures (e.g. [45], [6], [7], [11], [5]). The most frequently implemented theory is the OCC appraisal model ([30]), implemented in a number of systems and agents ([3], [1], [13]). Other emotion model implementations include models of emotions based on facial expression ([21] on recognition; [33] on synthesis), models of emotion based on blends of basic emotions ([33]), models of emotion based on Scherer's appraisal theory ([28]), models as goal management mechanisms ([16]), models of interaction of emotion and cognition ([2]), explicit models of the effects of emotion on cognitive processes ([20]), and effects of emotions on agent's belief generation ([17]).

Examples of integrated architectures focusing on emotion include most notably the work of Sloman and colleagues ([38]), but also more recent efforts to integrate emotion effects in Soar (a general cognitive architecture for developing systems that exhibit intelligent behaviour) by Jones and colleagues ([22]).

## 7    Is There a 'Right' Model for an Application?

As it is made clear throughout this chapter the choice of the emotion modelling approach for practical purposes does not have a one-phrase answer. In order to facilitate model choice, by pinpointing which emotion model is more fitting for which type of application, we propose a participatory design approach.

## 7.1    Participatory Design

Participatory design has been characterised as the 'third space' in Human Computer Interaction ([29]). In the world of software development, participatory design is an approach to design that attempts to involve the end users in the design process in a pro-active manner; this helps ensure that the outcome designed meets their needs and is usable. In participatory design, end-users (putative, potential or future) are invited to cooperate with researchers and developers during an innovation process. Potentially, they participate during several stages of an innovation process: they participate during the initial exploration and problem definition both to help define the problem and to focus ideas for solution, and during development, they help evaluate proposed solutions ([37]).

Participatory design can be seen as a way to move end-users views into the world of researchers and developers, whereas an empathic design approach would move researchers and developers into the world of end-users. Participatory design is expected to add an extra feedback loop in the showcase design phase. Each showcase designer will be able to put forward the detailed requirements that arise from corresponding use case scenarios. These use case scenarios can consequently be tested with the participation and active feedback of actual users. Such an approach is feasible thanks to their componential/modular structure. This structure allows for flexible decisions in the design process governed by the vital comments of end users. The suggested approach is similar to a formative evaluation of a system under development. It empowers users to engage in informed participation rather than being restricted to the use of existing, fully deployed and unchangeable systems.

This approach will help address open-ended and possibly multidisciplinary design problems that typically involve a combination of social and technological issues and don't have right or wrong answers.

## 7.2    Model Requirements vs. Application / Showcase Requirements

It has been made apparent throughout this text that the choice of the emotion modelling approach allows for flexible decisions and combinations of approaches in order to meet the showcase requirements. At this point we are going to identify some core requirements that make sense in all the showcases and then look into specific requirements that arise in each showcase separately.

Since we are talking about new media applications where interaction with the users is in the centre of attention we have to make sure it abides all basic usability restrictions. It is also crucial that the level of intrusiveness in the tracking process of emotion recognition is kept low. The ease of use of each showcase setup might conflict with the level of user engagement it manages to attain. It is obvious that the quality of the experience is prioritised highly and thus it is important to define the thresholds of acceptable intrusiveness conditions and complexity of setups in order to ensure it.

As underlined previously, context information during emotion modelling facilitates the process by providing knowledge about the event focus and the way it is appraised. Thus for each showcase the specific context is expected to dictate its own requirements towards the level of abstraction to be adopted, the deep or shallow modelling approach and the way temporal evolution of emotional measurements is dealt with. It would make sense to claim that a more detailed modelling approach is matching to the interactive-TV showcase whereas a more elementary modelling approach would correspond to a public space installation where the number of users and the environment noise constrain the level of detail of information collected.

# References

[1]    Andre, E., Klesen, M., Gebhard, P., Allen, S., Rist, T.: Integrating models of personality and emotions in lifelike characters. In: Paiva, A. (ed.) IWAI 1999. LNCS, vol. 1814, pp. 150–165. Springer, Heidelberg (2000)

[2]    Araujo, A.F.R.: Emotions influencing cognition. In: WAUME '93, Workshop on Architectures Underlying Motivation and Emotion, The University of Birmingham, Birmingham, UK (1993)

[3]    Bates, J., Loyall, A.B., Reilly, W.S.: Integrating reactivity, goals, and emotion in a broad agent. In: Proceedings of the 14th Meeting of the Cognitive Science Society, Boulder, CO (1992)

[4]    Breckler, S.J.: Empirical validation of affect, behaviour and cognition as distinct attitude components. Journal of Personality and Social Psychology 47(6), 1191–1205 (1984)

[5]    Breazeal, C.: Emotion and sociable humanoid robots. International Journal of Human-Computer Studies 59(1-2), 119–155 (2003)

[6]    Canamero, D.: Issues in the design of emotional agents. In: Proceedings of Emotional and Intelligent: The Tangled Knot of Cognition, AAAI Fall Symposium, TR FS-98-03, pp. 49–54. AAAI Press, Menlo Park (1998)

[7]    Castelfranchi, C.: Affective appraisal versus cognitive evaluations in social emotions and interactions. In: Paiva, A. (ed.) Affective Interactions: Towards a New Generation of Affective Interfaces, Springer, New York (2000)

[8]    Cowie, R., Douglas-Cowie, E., Apolloni, B., Taylor, J., Romano, A., Fellenz, W.: What a neural net needs to know about emotion words. In: Mastorakis, N. (ed.) Computational intelligence and applications, pp. 109–114. World Scientific Engineering Society (1999)

[9]    Cowie, R., Douglas-Cowie, E., Tsapatsoulis, N., Votsis, G., Kollias, S., Fellenz, W., Taylor, J.G.: Emotion recognition in human-computer interaction. IEEE Signal Processing Magazine 18, 32–80 (2001)

[10]   Damasio, A.: Looking for Spinoza: Joy, Sorrow, and the Feeling Brain. Harcourt Press, Orlando (2003)

[11] De Rosis, F., Pelachaud, C., Poggi, I., Carofiglio, V., Carolis, B.D.: From Greta's mind to her face: modeling the dynamics of affective states in a conversational embodied agent. International Journal of Human-Computer Studies 59(1-2), 81–118 (2003)

[12] Ekman, P.: An argument for basic emotions. Cognition & Emotion 6, 169–200 (1992)

[13] Elliott, C., Rickel, J., Lester, J.C.: Lifelike pedagogical agents and affective computing: An exploratory synthesis. In: Veloso, M.M., Wooldridge, M.J. (eds.) Artificial Intelligence Today. LNCS (LNAI), vol. 1600, pp. 195–211. Springer, Heidelberg (1999)

[14] Frijda, N.H.: The Emotions. Studies in Emotion and Social Interaction. Cambridge University Press, New York (1986)

[15] Frijda, N.H.: Emotions and Beliefs: How Feelings Influence Thoughts. Cambridge University Press, Cambridge (2000)

[16] Frijda, N.H., Swagerman, J.: Can computers feel? Theory and design of an emotional system. Cognition and Emotion 1(3), 235–257 (1987)

[17] Gratch, J., Marsella, S.: A domain-independent framework for modeling emotion. Cognitive Systems Research 5, 269–306 (2004)

[18] Goldie, P.: On Personality. Rutledge, New York (2004)

[19] Hudlicka, E.: To feel or not to feel: The role of affect in human-computer interaction. International Journal of Human-Computer Studies 59, 1–32 (2003)

[20] Hudlicka, E.: Increasing SIA architecture realism by modeling and adapting to affect and personality. In: Dautenhahn, K., Bond, A.H., Canamero, L., Edmonds, B. (eds.) Multiagent Systems, Artificial Societies, and Simulated Organizations, Kluwer Academic Publishers, Dordrecht (2002)

[21] Ioannou, S., Raouzaiou, A., Tzouvaras, V., Mailis, T., Karpouzis, K., Kollias, S.: Emotion recognition through facial expression analysis based on a neurofuzzy network. Neural Networks (Special Issue on Emotion: Understanding & Recognition) 18(4), 423–435 (2005)

[22] Jones, R., Henninger, A., Chown, E.: Interfacing emotional behavior moderators with intelligent synthetic forces. In: Proceedings of the 11th Conference on Computer Generated Forces and Behaviour Representation, Orlando, FL (2002)

[23] Karpouzis, K., Caridakis, G., Kessous, L., Amir, N., Raouzaiou, A., Malatesta, L., Kollias, S.: Modeling naturalistic affective states via facial, vocal, and bodily expressions recognition. In: Huang, T.S., Nijholt, A., Pantic, M., Pentland, A. (eds.) ICMI/IJCAI Workshops 2007. LNCS (LNAI), vol. 4451, pp. 91–112. Springer, Heidelberg (2007)

[24] Kant, I.: Critique of Judgment (1790) (Trans. Werner S. Pluhar). Hackett Publishing, Indianapolis (1987)

[25] Lazarus, R.S., Folkman, S.: Transactional theory and research on emotions and coping. European Journal of Personality 1, 141–169 (1987)

[26] Lazarus, R.S.: Emotion and Adaptation. Oxford University Press, New York (1991)

[27] Leventhal, H., Scherer, K.R.: The relationship of emotion to cognition: a functional approach to a semantic controversy. Cognition and Emotion 1, 3–28 (1987)

[28] Malatesta, L., Raouzaiou, A., Karpouzis, K., Kollias, S.: MPEG-4 facial expression synthesis. Personal and Ubiquitous Computing (Special issue on Emerging Multimodal Interfaces) 13(1), 77–83 (2007)

[29] Muller, M.J.: Participatory Design: The Third Space in HCI - Handbook of HCI. Erlbaum, Mahwah (2003)

[30] Ortony, A., Collins, A., Clore, G.L.: The Cognitive Structure of Emotions. Cambridge University Press, Cambridge (1988)

[31] Ortony, A., Turner, T.J.: What's basic about basic emotions? Psychological Review 97, 315–331 (1990)

[32] Picard, R.W.: Affective Computing. MIT Press, Cambridge (1997)

[33] Raouzaiou, A., Tsapatsoulis, N., Karpouzis, K., Kollias, S.: Parameterized facial expression synthesis based on MPEG-4. EURASIP Journal on Applied Signal Processing 2002(10), 1021–1038 (2002)

[34] Roseman, I.J.: The structure of emotion antecedents: Individual and cross-cultural differences. New approaches to emotion structure and process. Symposium conducted at the 92nd Annual Convention, American Psychological Association, Toronto, Canada (1984)

[35] Roseman, I.J.: The emotion system: Strategies for coping with crises and opportunities. Paper presented at Department of Psychology, The Graduate Center, City University of New York (1990)

[36] Rusting, C.: Personality, mood, and cognitive processing of Emotional information: three conceptual frameworks. Psychological Bulletin 124, 165–196 (1998)

[37] Schuler, D., Namioka, A. (eds.): Participatory Design: Principles and Practices. Lawrence Erlbaum Associates, Hillsdale (1993)

[38] Sloman, A.: Beyond shallow models of emotions. Cognitive Processing 2(1), 177–198 (2001)

[39] Smith, C.A., Kirby, L.D.: Affect and appraisal. In: Forgas, J.P. (ed.) Feeling and Thinking: The Role of Affect in Social Cognition, Cambridge University Press, Cambridge (2000)

[40] Scherer, K.R.: Toward a dynamic theory of emotion: The component process model of affective states. Geneva Studies in Emotion and Communication 1, 1–98 (1987)

[41] Scherer, K.R.: Criteria for emotion-antecedent appraisal: A review. In: Hamilton, V., Bower, G.H., Frijda, N.H. (eds.) Cognitive perspectives on emotion and motivation, pp. 89–126. Kluwer Academic Publishers, Dordrecht (1988)

[42] Scherer, K.R.: Appraisal considered as a process of multi-level sequential checking. In: Scherer, K.R., Schorr, A., Johnstone, T. (eds.) Appraisal processes in emotion: Theory, methods, research, pp. 92–120. Oxford University Press, New York (2001)

[43] Scherer, K.R.: Feelings integrate the central representation of appraisal-driven response organization in emotion. In: Manstead, A.S.R., Frijda, N.H., Fischer, A.H. (eds.) Feelings and emotions The Amsterdam symposium, pp. 136–157. Cambridge University Press, Cambridge (2004)

[44] Scherer, K.R.: Unconscious processes in emotion: The bulk of the iceberg. In: Niedenthal, P., Feldman-Barrett, L., Winkielman, P. (eds.) The unconscious in emotion, Guilford, New York (2005)

[45] Velasquez, J.: Modeling Emotions and Other Motivations in Synthetic Agents. In: Proceedings of AAAI-97, Providence, RI, pp. 10–15 (1997)

# Introducing Intelligence in Electronic Healthcare Systems: State of the Art and Future Trends

Ilias Maglogiannis

University of Central Greece, Papasiopoulou 2-4, PC 35100 Lamia, Greece
imaglo@ucg.gr

**Abstract.** This chapter introduces intelligent technologies applied in electronic healthcare systems and services. It presents an overview of healthcare technologies that enable the advanced patient data acquisition and management of medical information in electronic health records. The chapter presents the most important patient data classification methods, while special focus is placed on new concepts in intelligent healthcare platforms (i.e., advanced data mining, agents and context-aware systems) that provide enhanced means of medical data interpretation and manipulation. The chapter is concluded with the areas in which intelligent electronic healthcare systems are anticipated to make a difference in the near future.

## 1    Introduction

In this era of ubiquitous and mobile computing the vision in biomedical informatics is towards achieving two specific goals: the availability of software applications and medical information anywhere and anytime and the invisibility of computing [26]. Both these goals lead to the introduction of electronic healthcare computing concepts and features in e-health applications. Applications and interfaces that will be able to automatically process data provided by medical devices and sensors, exchange knowledge and make intelligent decisions in a given context are strongly desirable. Natural user interactions with such applications are based on autonomy, avoiding the need for the user to control every action, and adaptivity, so that they are contextualized and personalized, delivering the right information and decision at the right moment [27]. All the above recently introduced features provide added value in modern electronic healthcare systems.

These technologies can support a wide range of applications and services including automated diagnosis, personalized medicine, patient monitoring, location-based medical services, emergency response and management, ubiquitous access to medical data, and home monitoring. This chapter presents a special branch of artificial intelligence tools and applications called intelligent electronic healthcare systems. In general, the term intelligent electronic healthcare systems refers to automated systems that process medical data such as clinical examinations or

M. Bramer (Ed.): Artificial Intelligence, LNAI 5640, pp. 71–90, 2009.

medical images and provide estimated diagnoses. The estimations are often based on the analysis of details that elude the human eye as well as large amounts of medical history that humans cannot possibly consider or analyzing non-visual characteristics of medical data. Although such systems typically do not reach 100% success, which means that they cannot substitute the working physician, the input they provide is extremely helpful as an independent source of evidence concerning a correct medical decision.

The development of intelligent health-care systems is a very promising area for commercial organizations active in the health monitoring domain. Currently, the cost effective provision of quality healthcare is a very important issue throughout the world since healthcare faces a significant funding crisis due to the increasing population of older people and the reappearance of diseases that should be controllable. Intelligent healthcare systems are capable of attacking all these challenges in an efficient and cost-effective way. Hardware and software is gradually becoming cost-affordable, can be installed and operated in numerous sites (frequently visited by patients), can be interfaced to a wide variety of medical information systems (e.g., patient databases, medical archives), thus involving numerous actors. Hence, the electronic health systems in general present a truly scalable architecture covering a wide spectrum of business roles and models [23].

This chapter aims at presenting the state of the art and new trends in intelligent healthcare systems. The chapter is structured as follows: Section 2 discusses the technologies that enable the use of healthcare computing (i.e., patient data acquisition methods and tools, medical data management, healthcare information systems and medical data exchange). Section 3 overviews the intelligent aspect that can be applied in electronic healthcare systems, while Section 4 focuses on new concepts in electronic healthcare applications such as intelligent agents and context-awareness and finally, Section 5 presents the challenges of the near future and concludes this chapter.

# 2    HealthCare Enabling Technologies

## 2.1    Patient Biosignals and Acquisition Methods

A broad definition of a signal is a 'measurable indication or representation of an actual phenomenon', which in the field of biosignals, refers to observable facts or stimuli of biological systems or life forms. In order to extract and document the meaning or the cause of a signal, a physician may utilize simple examination procedures, such as measuring the temperature of a human body or may have to resort to highly specialized and sometimes intrusive equipment, such as an endoscope. Following signal acquisition, physicians go on to a second step, that of interpreting its meaning, usually after some kind of signal enhancement or 'pre-processing', that separates the captured information from noise and prepares it for specialized processing, classification and decision support algorithms.

Biosignals require a digitization step in order to be converted into a digital form. This process begins with acquiring the raw signal in its analog form, which is then fed into an analog-to-digital (A/D) converter. Since computers cannot handle or store continuous data, the first step of the conversion procedure is to produce a discrete-time series from the analog form of the raw signal. This step is known as 'sampling' and is meant to create a sequence of values sampled from the original analog signals at predefined intervals, which can faithfully reconstruct the initial signal waveform. The second step of the digitization process is quantization, which works on the temporally sampled values of the initial signal and produces a signal, which is both temporally and quantitatively discrete; this means that the initial values are converted and encoded according to properties such as bit allocation and value range. Essentially, quantization maps the sampled signal into a range of values that is both compact and efficient for algorithms to work with. The most popular biosignals utilized in electronic healthcare applications ([1], [3], [4], [10], [11], [16], [17], [19], [20], [23]) are summarized in Table 1.

**Table 1.** Broadly used biosignals with corresponding metric ranges, number of sensors required and information rate.

| Biomedical Measurements (Broadly Used Biosignals) | Voltage range (V) | Number of sensors | Information rate (b/s) |
|---|---|---|---|
| ECG | 0.5-4 m | 5-9 | 15000 |
| Heart sound | Extremely small | 2-4 | 120000 |
| Heart rate | 0.5-4 m | 2 | 600 |
| EEG | 2-200 μ | 20 | 4200 |
| EMG | 0.1-5 m | 2+ | 600000 |
| Respiratory rate | Small | 1 | 800 |
| Temperature of body | 0-100 m | 1+ | 80 |

In addition to the aforementioned biosignals, patient physiological data (e,g., body movement information based on accelerometer values), and context-aware data (e.g., location, environment and age group information) have also been used by electronic healthcare applications ([1], [2], [3], [4], [6], [13], [14], [15], [18], [20], [21], [24]). The utilization of the latter information is discussed in the following sections.

In the context of healthcare applications, the acquisition of biomedical signals is performed through special devices (i.e. sensors) attached on the patients body (see Fig. 1) or special wearable devices (see Fig. 2). Regarding the contextual information, most applications are based on data collected from video cameras, microphones, movement and vibration sensors.

**Fig. 1.** Accelerometer sensor device that can be attached on patient's body and transmit movement data wirelessly to the monitoring unit [21].

**Fig. 2.** CodeBlue [9]: A wearable ECG and pulse oximeter measurement device.

## 2.2    Healthcare Information Systems and Medical Data Exchange

The use of healthcare information systems and potential applications are numerous nowadays. Medical platforms allowing doctors to access Electronic Health Records (EHR) are already set up in several countries [33], [34], [35]. An EHR is an electronic version of a patient's medical history, that is maintained by the healthcare provider over time, and includes all of the key administrative clinical data relevant to that person's care under a particular provider, including demographics, progress notes, problems, medications, vital signs, past medical history, immunizations, laboratory data, medical images and radiology reports.

The EHR automates access to information and has the potential to streamline the clinician's workflow. The EHR also has the ability to support other care-related activities directly or indirectly through various interfaces, including evidence-based decision support, quality management, and outcomes reporting. The type of data included in EHR systems are presented in Table 2.

Table 2. Electronic Health Records (EHR) data modalities

| Digital Data | Contrast / Resolution (No. of samples per second x bits per sample) | Data Size |
|---|---|---|
| Demographic Data | | ~ 100 KB |
| Clinical Data (Biosignals) | | ~ 100 KB / incident |
| Digital audio stethoscope (Heart Sound) | 10000 x 12 | ~ 120 kbps |
| Electrocardiogram ECG | 1250 x 12 | ~ 15 Kbps |
| Electroencephalogram EEG | 350 x 12 | ~ 10 Kbps |
| Electromyogram EMG | 50000 x 12 | ~ 600 Kbps |
| Ultrasound, Cardiology, Radiology | 512x512x8 | 256 KB (image size) |
| Magnetic resonance image | 512x512x8 | 384 KB (image size) |
| Scanned x-ray | 1024x1250x12 | 1.8 MB (image size) |
| Digital radiography | 2048x2048x12 | 6 MB (image size) |
| Mammogram | 4096x4096x12 | 24 MB (image size) |
| Compressed and full motion video (telemedicine) | - | 384 kbps to 1.544 Mb/s (speed) |

EHR systems provide the hospitals with the infrastructure to collaborate efficiently at a technical level. Hospitals are sufficiently rich in their infrastructure to handle the internal administrative and clinical processes and the need to integrate the processes of geographically distributed and organizationally independent organizations is evident. At business level, however, the need to integrate the processes of geographically distributed and organizationally independent organizations led the design of architecture of health information systems to combine the principles of different approaches to interoperability: Workflow Management Systems (WfMSs), the Middleware approaches to interoperability such as Message Oriented Middleware, the Semantic Web and Visual Integration. A brief reference to the above approaches is given in the following paragraphs.

Workflow is defined [36] as the computerized facilitation or automation of a business process, in whole or part and a Workflow Management System is a system that completely defines, manages and executes "workflows" through the execution of software whose order of execution is driven by a computer representation of the workflow logic. Interoperability among workflow products concerns a standardized set of interfaces and data interchange formats between such components in the health care sector.

### 2.2.1   The Semantic Web – Web Services

The World Wide Web was initially designed for unstructured information exchange, but that led to lack of uniformity for accessing web services. To facilitate access to complex services, a group of companies standardized on SOAP (Simple Object Access Protocol) as a light-weight protocol based on XML for exchanging messages over the Web. Similarly higher-level service layers have been defined such as WSDL (Web Service Description Language) and UDDI (Universal Description, Discovery and Integration). The use of ontology is suggested and languages for specification and representation of knowledge in the semantic web like OWL, OIL, DAML+OIL, UDDI are used.

### 2.2.2   Message Oriented Standards – HL7

Health Level Seven, Inc. ([37]) is a not-for-profit, ANSI-Accredited Standards Developing Organization that provides standards for the exchange, management and integration of data that supports clinical patient care and the management, delivery and evaluation of healthcare services. Data exchange is implemented by exchanging messages. HL7 corresponds to the conceptual definition of an application-to-application interface placed in the 7th layer of the OSI model. HL7 achieves interoperability through syntactically and semantically standardized messages. In the US and in many European countries the HL7 standard has become the main communication standard for healthcare system integration.

### 2.2.3   Message Oriented Standards – CEN/TC 251 Health Informatics

CEN is a European collaboration of the formal standards bodies of 19 countries with strong links to the politics of the European Union and with Eastern European countries as associate members. The standardization of Health Informatics started in 1990 and has resulted in a number of message standards based on information models, most often implemented in Edifact, but since 1999, also implementable in XML. The standards work of CEN/TC 251 complements HL7 work in the areas of security, healthcare record architecture and device communication. The standard that has been defined for the field of health informatics is a messaging standard and also provides the architecture concept for the middleware layer for healthcare-specific applications.

### 2.2.4   Message Oriented Standards – DICOM

The great majority of equipment that deals with digital medical imaging and communication supports DICOM. It supports operation in a networked environment using the industry standard networking protocol TCP/IP. The standard specifies how devices react to commands and data being exchanged. The creation of

DICOM Structured Reporting (DICOM-SR), in the year 2000, has established a method for constructing and transferring information objects that encode structured documents. Structured reports ease the search for specific information, report translation and comparison between different findings. The standard explicitly describes how reports are structured, using controlled terminology like SNOMED.

### 2.2.5 CORBA

Distributed application frameworks required to build complex services have been around for a while. Popular ones are (or have been) COM (Component Object Model), DCOM (Distributed COM), and COM+ which are Microsoft specific, EJB (Enterprise Java Beans) which is Java specific, and CORBA (Common Object Request Broker Architecture) [37] which is both platform and language independent. CORBA is created and maintained by the Object Management Group (OMG), an international, non-profit software organization driven and supported by information system vendors' software developers and technology users. To address the needs of the rapidly changing healthcare industry, the OMG established a Healthcare Task Force, the CORBAmed. A key difference between CORBA and Web Service Technologies (UDDI/WSDL/SOAP) is that CORBA provides true object-oriented component architecture unlike the Web services, which are primarily message–based [38]. Moreover CORBA also comes with a standard set of services (Events, Naming, Trading) that allow application writers to focus on the business logic rather than on the details of the communication infrastructure.

The above mentioned standards enable the interoperability of electronic healthcare systems and in addition facilitate the collection of large medical datasets describing logical organization of same or similar pathological conditions for one or many patients. These medical datasets are the basis for the development of intelligent systems, allowing the advanced processing and interpretation of physiological, clinical and image medical data. These systems are encountered mostly as advanced Medical Decision Support Systems (MDSS) that could help medical professionals as diagnostic adjuncts promoting the quality of medical services, especially in underserved populations, where expert medical knowledge is unavailable. This aspect of electronic healthcare systems is discussed in the next section.

## 3    Artificial Intelligence in Electronic Healthcare Systems

The objective of computer-assisted decision making in healthcare aims to allow the medical professional to use the computer as a tool in the decision process. The most important processes in the development and operation of Medical Decision Support Systems (MDSS) are (i) the acquisition of information regarding the diagnosis classes and (ii) the actual classification of a given case to a diagnosis. The two steps are actually closely related to each other, as the type of classifier chosen in most cases also indicates the training methodology to use. Although there is an extremely wide variety of classification methodologies that one may choose to apply, the most well known and widely applied genres of approaches can be briefly summarized and categorized as follows:

1. The examined case is compared directly to other cases in the EHR and similarities are used in order to provide a most probable diagnosis.
2. Different types of classifiers are trained based on available health records, so that the underlying data patterns are automatically identified and utilized in order to provide more reliable classification of future data.
3. Information extracted automatically from medical history, or provided manually by human experts, is organized so that diagnosis estimation is provided in a dialogical manner through a series of question/answer sessions.
4. Multiple simple classifiers are combined in order to minimize the error margin.
5. Information provide by human experts in the form of simple rules is utilized by a fuzzy systems in order to evaluate the case in hand.

The following subsections discuss briefly the data classification methods used in MDSS and the corresponding evaluation methodologies.

### 3.1     Patient Data Classification Methods

Data classification is an important problem in a variety of engineering and scientific disciplines such biology, psychology, medicine, marketing, computer vision, and artificial intelligence [30]. Its main object is to classify objects into a number of categories or classes. Depending on the application, these objects can be images or signal waveforms or any type of measurements that need to be classified. Given a specific data feature, its classification may consist of one of the following two tasks: a) supervised classification in which the input pattern is identified as a member of a predefined class; b) unsupervised classification in which the pattern is assigned to a hitherto unknown class.

In statistical data classification, input data are represented by a set of n features, or attributes, viewed as an n-dimensional feature vector. The classification system is operated in two modes: training and classification. Data preprocessing can be also performed in order to segment the pattern of interest from the background, remove noise, normalize the pattern, and any other operation which will contribute in defining a compact representation of the pattern. In the training mode, the feature extraction/selection module finds the appropriate features for representing the input patterns and the classifier is trained to partition the feature space. The feedback path allows a designer to optimize the preprocessing and feature extraction/selection strategies. In the classification mode, the trained classifier assigns the input pattern to one of the pattern classes under consideration based on the measured features.

There is a vast array of established classification techniques, ranging from classical statistical methods, such as linear and logistic regression, to neural network and tree-based techniques. In the following we review the main categories of classification systems that find application in an MDSS framework.

### 3.1.1  k- Nearest Neighbours

The $k$-Nearest Neighbours methodology, often referred to as $k$-NN, constitutes a breed of classifiers that attempt to classify the given patient data by identifying other similar cases in his or other health records. The simplest, as well as most common, case is when all considered features extracted by medical data are scalars. The $k$-NN methodology has found many applications in the field of MDSSs [44], [45], [46].

### 3.1.2  Artificial Neural Networks

Artificial Neural Networks (ANNs) are the main representatives of a more robust approach to classification; these classifiers, prior to being put to use, process available medical data in EHRs in order to extract useful information concerning the underlying data patterns and structure, thus also acquiring the information required in order to optimize the classification process. Following a massively parallel architecture, quite similar to that of neurons in the human brain, ANNs construct powerful processing and decision making engines through the combination of quite trivial processing units – also named neurons. ANNs are exhaustively used in MDSS. For instance, the following works are based on applications of ANNs [47], [48], [49], [50], while [51] reviews the benefits of ANN application in medicine in general.

The characteristic that makes ANNs so popular is the fact that given a set of labeled data (extracted by the EHR) an ANN will tune itself in an automated manner so as to match these data in the best possible way. Unfortunately, this training process is not a trivial one. Numerous methodologies are presented in the literature for training ANNs, each one focusing on a different feature or situation. Thus, different training methodologies (as well as network structure) can be used when the amount of training data is small, computing resources are limited, some or all of the medical data are unlabelled. Therefore the adoption of the standards discussed in section 2 is considered a necessity.

### 3.1.3  Self Organizing Maps

The fact that one needs to have a clear idea concerning the structure of the network (count of layers, count of neurons) before even training and testing can start is a very important limitation for ANNs. Kohonen's Self Organizing Maps (SOMs) constitute a more interesting and robust approach to training a network with no prior knowledge of the underlying data structures. A number of details about the selection of the parameters, variants of the map, and many other aspects have been covered in the monograph [52]. Due to their excellent properties and characteristics, SOMs have found numerous applications in MDSSs, such as [53], [54], [55], [56].

### 3.1.4  Support Vector Machines

The Support Vector Machine (SVM) is a novel algorithm for data classification and regression. It was introduced by Vapnic and is clearly connected with statistical learning theory [57], [58], [59]. The SVM is an estimation algorithm that separates data in two classes, but since all classification problems can be restricted to consideration of the two-class classification problem without loss of generality, SVMs can be applied to classification problems in general. SVMs allow the expansion of the information provided by a training data set as a linear combination of a subset of the data in the training set (support vectors). These vectors locate a hypersurface that separates the input data with a very good degree of generalization. The SVM algorithm is a learning machine; therefore it is based on training, testing and performance evaluation, which are common steps in every learning procedure. Training involves optimization of a convex cost function where there are no local minima to complicate the learning process. Testing is based on the model evaluation using the support vectors to classify a test data set. Performance is based on error rate determination as test set data size tends to infinity. Due to the fact that SVMs focus on maximizing the margin between classes, thus minimizing the probability of misclassification, they are extremely popular in MDSSs, where the cost of a misclassification may have a direct impact on human life. The following works are just a few examples of works in the medical field that are based on SVM learning [60], [61], [62], [63], [64].

### 3.1.5  Decision Trees

Physicians using MDSSs are often reluctant to leave important medical decisions to a sub-symbolic, and thus generally incomprehensible, automated engine. Decision trees offer an alternative computing methodology which reaches a decision through consecutive, simple question and answer sessions.

In the learning phase (when the decision tree is constructed) exactly one of the available features needs to be selected as the root feature, i.e. the most important feature in determining the diagnosis. Then data are split according to the value they have for this feature, and each group of data is used in order to create the corresponding child (sub-tree) of the root. If all of the data in a group belongs to the same diagnosis, then that child becomes a leaf to the tree and is assigned that diagnosis. Otherwise, another feature is selected for that group, and data are again split leading to new groups and new children for this node. Decision trees are also widely used for the development of MDSS [65], [66], [67], [68], [69]. A review of decision tree applications in medicine is available in [70].

## 3.2     Performance Evaluation of Classification Systems

The performance of each classifier is tested using an ideally large set of manually classified data. A subset of them, e.g., 80% is used as the training set and the remaining 20% of the samples are used for testing using the trained classifier. The training and test data are exchanged for all possible combinations to avoid bias in the solution. Classification performance of MDSS is typically based on a true/false and positive/negative scheme. When adopted in the medical case, true positive (TP) is correct illness estimation, true negative (TN) a correct healthy estimation, false positive (FP) illness estimation for a healthy case and a false negative (FN) a healthy estimation for an ill case. Based on these, accuracy is defined as follows:

$$Accuracy = \frac{TP+TN}{TP+TN+FP+FN} \qquad (Equation\ 1)$$

The simplistic approach of simply counting correct and incorrect classifications in order to estimate accuracy, although generally accepted in other expert systems and classifiers, is not sufficient for the case of medical systems, where one type of mistake may be much more important – as far as the possible consequences are concerned – compared to another. For example a false positive estimation has the result of a patient taking extra tests in order to verify their health status, whereas a false negative diagnosis may deprive them of early diagnosis and treatment. Finally, classes in a medical setting are rarely balanced; it is typical that only a small percentage of people examined will actually be ill. As a result, a system that always provides a "healthy" diagnosis reaches high classification rates.

In order to compensate for this, a more flexible consideration of errors needs to be used, in order for class probabilities to be considered as well. A simple approach that is commonly followed in this direction is the utilization of specificity and sensitivity measures, defined as follows:

$$Specificity = \frac{TN}{TN+FP} \qquad (Equation\ 2)$$

$$Sensitivity = \frac{TP}{TP+FN} \qquad (Equation\ 3)$$

where specificity and sensitivity are actually measures of accuracy, when considering only healthy or only ill cases, respectively, thus decoupling the measures from class probabilities.

A graphical representation of classification performance is the Receiver Operating Characteristic (ROC) curve (see Fig. 3), which displays the "tradeoff" between sensitivity (i.e. TPF) and specificity (i.e. TNF) that results from the overlap between the distribution of lesion scores for ill and healthy data. A good classifier is one with close to 100% sensitivity at a threshold such that high specificity is also obtained. The ROC for such a classifier will plot as a steeply rising curve. When different classifiers are compared, the one whose curve rises fastest should be best. If sensitivity and specificity were weighted equally, the greater the area under the ROC curve (AUC), the better the classifier. An extension of ROC analysis found in the literature [39] is the three-way ROC analysis that applies to trichotomous tests. It summarizes the discriminatory power of a trichotomous test in a single value, called the volume under surface (VUS) by analogy to the AUC value for dichotomous tests. Just as the AUC value for dichotomous tests is equivalent to the probability of correctly ranking a given pair of normal and abnormal cases, the VUS value for trichotomous tests is equivalent to the probability of correctly distinguishing three cases, where each case is from a different class.

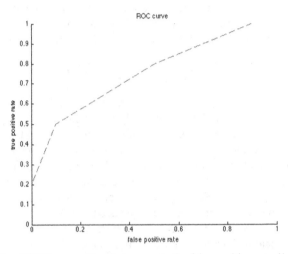

**Fig. 3.** Example of ROC curve. X-axis represents the false positive rate (1-Sp, where Sp is the specificity) and the Y-axis the true positive rate (or Sensitivity, Se).

# 4     New Concepts: Intelligent Agents and Context Awareness

As can be seen from the above section, several research efforts dealing with machine intelligence techniques on clinician settings providing advanced healthcare services exist in the literature. All of the surveyed works present corresponding clinical trials of the effects and patient outcomes from the application of such Medical Decision Support Systems (MDSS). State of the art works in the field of intelligent electronic healthcare systems, report that the new concepts and approaches deal with advanced data mining and intelligent agents, while context awareness is the new desirable feature of e-health applications. The next two subsections analyze the aforementioned newly introduced approaches.

## 4.1     Data Mining and Intelligent Agents

The proliferation of healthcare data has resulted in a large number of efforts to inductively manipulate, interpret and discover 'useful' knowledge from the collected data. Interesting results have been reported by health informatics research-ers using a variety of advanced Data Mining (DM) algorithms [31], [32].

The most important anticipated tasks for the medical repositories are summa-rized into the following:

- Problem analysis and specification, which guides the choice of 'appropriate' DM

- Establishing a communication channel to enable remote access to the data re-positories of multiple hospitals. Technically this involves the exchange of mes-sages.

- Collection of 'relevant' data to complete each individual task need to be first identified and subsequently retrieved from the respective data repositories.

- Synthesis of heterogeneous data originating from multiple data repositories.

- Preparation of the data according to the specification of the DM service packages.

- Execution of the DM algorithm.

- Generation of a DM report for the end-user.

Due to the existence of multiple heterogeneous data repositories in a healthcare enterprise, a distributed data community should be established, such that any DM effort draws upon the 'holistic' data available within the entire healthcare enter-prise. Multi Agent-Based Data Mining Info-Structures (ADMI), responsible for the generation of data-mediated diagnostic-support and strategic services have been proposed. The latter takes advantage of a multi-agent architecture, which features the amalgamation of various types of intelligent agents.

Intelligent agents can be viewed as autonomous software (or hardware) con-structs that are proactively involved in achieving a predetermined task and at the same time reacting to its environment. According to [29], agents are capable of:

- performing tasks (on behalf of users or other agents).

- interacting with users to receive instructions and give responses.

- operating autonomously without direct intervention by users, including moni-toring the environment and acting upon the environment to bring about changes.

- showing intelligence – to interpret monitored events and make appropriate decisions.

Agents can be proactive, in terms of being able to exhibit goal-directed behavior, reactive; being able to respond to changes in the environment, including detecting and communicating to other agents, autonomous; making decisions and control-ling their actions independently of others. Intelligent agents can be also considered as social entities where they can communicate with other agents using an agent-communication language in the process of carrying out their tasks. Software agents can also be used in order to perform distributed analysis of vital data and give an alarm indication to previously-selected physicians and family members [11]. Agents may also assist patients or treatment experts to perform basic tasks like meal preparation and medication [11], [12].

Additional Agent-based techniques [7] can often be utilized for modeling application components as somewhat autonomous agents that easily reflect health-care institutions' decentralized networks. Medical agent interfaces ([5], [23]) provide continuous and more direct access to the aforementioned information. Software agents are installed either on mobile devices (e.g., PDAs) or on interactive devices within the treatment center (e.g. PCs or LCD monitors, or smart walls [11]). Information retrieval and presentation can be either performed by user request or reactively (e.g, based on user's location or patient's state). Queries regarding patient data or medical information (e.g., medication procedures, diseases symptoms, etc.) are parsed through specific agents (i.e. query optimization agents) and forwarded to knowledge retrieval agents for research. The information retrieval can be performed either from the local hospital information system or remote medical knowledge repositories. Information retrieval, knowledge adaptation and presentation to the user are performed by related agents using medical ontologies for proper knowledge data representation [8].

Using such advanced knowledge representation and medical data retrieval methods, to access multiple healthcare information is feasible, even from mobile devices. Proper access restriction to sensitive information can be applied and direct access to important information in cases of emergency can be established [25].

## 4.2    Context Awareness

Context awareness is the capability of ehealth applications to be aware of the existence and characteristics of the patient's activities and environments. In rapidly changing scenarios, such as the ones considered in the fields of biomedical informatics, systems have to adapt their behaviour based on the current conditions and the dynamicity of the environment they are immersed in ([28]). A system is context-aware if it can extract, interpret and use context information and adapt its functionality to the current context of use. The challenge for such systems lies in the complexity of capturing, representing and processing contextual data. To capture context information generally some additional sensors and/or programs are required [22]. The main goal of context aware computing is to acquire and utilize information about the context of a medical device to provide services that are appropriate to particular people, place, time, events, etc. ([42]). According to the latter, the work presented in [40] describes a context-aware mobile system for inter-hospital communication taking into account patient's and physician's physical location for instant and efficient messaging regarding medical events. J. Bardram presents in [41] additional cases of context-awareness used within treatment centres and provides design principles for such systems. The project 'AWARENESS' (presented in [43]) provides a more general framework for enhanced telemedicine and telediagnosis services depending on patient status and location.

The way context-aware applications make use of context can be categorized into the three following classes: presenting information and services, executing a service, and tagging captured data.

*Presenting information and services* refers to applications that either present context information to the user, or use context to propose appropriate selections of actions to the user.

*Automatically executing a service* describes applications that trigger a command, or reconfigure the system on behalf of the user according to context changes.

*Attaching context information for later retrieval* refers to applications that tag captured data with relevant context information.

The patient state can be determined through a number of biosensors (i.e. heart rate and body temperature sensors) and corresponding vital signals. Defined threshold values in the latter signals determine the case of an immediate medical data transmission (alarm event) to the monitoring unit. In case of normal patient status, periodical summarized data transmission might occur at lower detail. Data coding and transmission can also vary according to network availability and quality: Context awareness can be used for instance in cases of remote assessment or telesurgery. According to the network interface used, appropriate video coding is applied to the transmitted medical data, avoiding thus possible transmission delays and optimizing a telemedicine procedure.

## 5    Discussion and Conclusions

As clinical machine intelligence techniques mature, it seems they can offer increasingly exciting prospects for improving the effectiveness and efficiency of patient care and the development of more reliable intelligent electronic healthcare systems. According to a recent review [71] published studies of clinical machine intelligence systems are increasing rapidly, and their quality is improving. It seems that they may enhance clinical performance for drug studies, preventive care, and other aspects of medical care, but not convincingly however in all cases for diagnosis and prognosis. The potential reason for this is that rigorous evaluations of MDSSs are usually more difficult to conduct than evaluations of drug studies, for instance, because clinical settings often preclude complete separation of the intervention and control groups. The studies of patient outcomes require also large numbers of participants and significant budgets, which are not always easy to find. Without the existence of such rigorous patient outcomes studies physicians may not be convinced to introduce the use of MDSSs in the routine practice of healthcare.

Clearly, the goal is to reach a stage where intelligent electronic healthcare systems are integrated in the process of everyday clinical work, but without being assigned roles they are not made for, such as the role of the actual clinician. It seems that a number of parameters will have an effect in this process, ranging from purely financial issues to degree of automation, from availability at the time and location it is needed to the ease of the user interface and from the adoption of standards in medical data acquisition components to the success of the system development and integration procedures. The areas in which intelligent electronic healthcare systems could make a difference are many. The following Table provides a summary of the most important ones in the author's view.

**Table 3.** Potential users and uses for intelligent electronic healthcare systems

| User | Application |
| --- | --- |
| Pharmacists | Drug levels, drug/drug interactions, culture & sensitivity results, adverse drug events |
| Physicians | Advanced Medical Data Processing Tools, Extraction of Features, Quantification of Pathological Phenomena |
| Non-Expert Physician | Computer Supported Diagnosis |
| Remote Physician | Advanced Telemedicine Systems |
| Biologists | Simulation of pathogenetic mechanisms |
| Nurses | Critical lab results, drug/drug interactions |
| Dietary | Patient transfers, lab support for tube feedings |
| Epidemiology/infection control | Epidemiological results, reportable organisms |
| Homecare | Patient Monitoring at Home |
| Billing | Excessively expensive tests and treatments |
| Administration | Patient chart administration |
| Patient | Drug/drug interactions, drug dosing, missing tests |

# References

[1] Corchado, J.M., Bajo, J., de Paz, Y., Tapia, D.I.: Intelligent Environment for monitoring Alzheimer patients, agent technology for healthcare. To be published in Decision Support Systems, article available online at http://www.sciencedirect.com

[2] Sharmin, M., Ahmed, S., Ahamed, S.I., Haque, M.M., Khan, A.J.: Healthcare aide: towards a virtual assistant for doctors using pervasive middleware. In: Proc. of Fourth Annual IEEE International Conference on Pervasive Computing and Communications Workshops, pp. 6–12 (2006)

[3] Paganelli, F., Spinicci, E., Mamelli, A., Bernazzani, R., Barone, P.: ERMHAN: A multi-channel context-aware platform to support mobile caregivers in continuous care networks. In: Proc. of IEEE International Conference in Pervasive Technologies, pp. 355–360 (2007)

[4] Mitchell, S., Spiteri, M.D., Bates, J., Coulouris, G.: Context-Aware Multimedia Computing in the Intelligent Hospital. In: Proc. SIGOPS EW2000, the Ninth ACM SIGOPS European Workshop (2000)

[5] Hashmi, Z.I., Abidi, S.S.R., Cheah, Y.-N.: An Intelligent Agent-based Knowledge Broker for Enterprise-wide Healthcare Knowledge Procurement. In: 15th IEEE Symposium on Computer-Based Medical Systems (CBMS'02), p. 173 (2002)

[6]  Choudhri, A., Kagal, L., Joshi, A., Finin, T., Yesha, Y.: PatientService: Electronic
     Patient Record Redaction and Delivery in Pervasive Environments. In: Fifth Interna-
     tional Workshop on Enterprise Networking and Computing in Healthcare Industry
     (2003)
[7]  Kifor, T., Varga, L., Vazquez-Salceda, J., Alvarez, S., Miles, S., Moreau, L.: Prove-
     nance in Agent-Mediated Healthcare Systems. IEEE Intelligent Systems 21(6), 38–46
     (2006)
[8]  Moreno, A., Valls, A., Isern, D., Sanchez, D.: Applying Agent Technology to Health-
     care: The GruSMA Experience. IEEE Intelligent Systems 21(6), 63–67 (2006)
[9]  Malan, D., Fulford-Jones, T., Welsh, M., Moulton, S.: CodeBlue: An Ad Hoc Sensor
     Network Infrastructure for Emergency Medical Care. In: International Workshop on
     Wearable and Implantable Body Sensor Networks (2004)
[10] Gouaux, F., Simon-Chautemps, L., Adami, S., Arzi, M., Assanelli, D., Fayn, J., Forlini,
     M.C., Malossi, C., Martinez, A., Placide, J., Ziliani, G.L., Rubel, P.: Smart devices for
     the early detection and interpretation of cardiological syndromes. In: 4th International
     IEEE EMBS Special Topic Conference on Information Technology Applications in
     Biomedicine, pp. 291–294 (2003)
[11] Jeen, Y., Kim, J., Park, J., Park, P.: Design and implementation of the Smart Health-
     care Frame Based on Pervasive Computing Technology. In: The 9th International
     Conference on Advanced Communication Technology, pp. 349–352 (2007)
[12] Camarinha-Matos, L.M., Vieira, W.: Intelligent mobile agents in elderly care. Robot-
     ics and Autonomous Systems 27, 59–75 (1999)
[13] Barger, T.S., Brown, D.E., Alwan, M.: Health-Status Monitoring Through Analysis of
     Behavioral Patterns. IEEE Transactions on Systems, Man and Cybernetics 35(1), 22–
     27 (2005)
[14] Starida, K., Ganiatsas, G., Fotiadis, D.I., Likas, A.: CHILDCARE: a collaborative
     environment for the monitoring of children healthcare at home. In: 4th International
     IEEE EMBS Special Topic Conference on Information Technology Applications in
     Biomedicine, pp. 169–172 (2003)
[15] Jansen, B., Deklerck, R.: Context aware inactivity recognition for visual fall detection.
     In: Pervasive Health Conference and Workshops, pp. 1–4 (2006)
[16] Jannett, T.C., Prashanth, S., Mishra, S., Ved, V., Mangalvedhekar, A., Deshpande, J.:
     An intelligent telemedicine system for remote spirometric monitoring. In: Proceedings
     of the Thirty-Fourth Southeastern Symposium on System Theory, pp. 53–56 (2002)
[17] Dolgov, A.B., Zane, R.: Low-Power Wireless Medical Sensor Platform. In: 28th
     Annual International Conference of the IEEE Engineering in Medicine and Biology
     Society, pp. 2067–2070 (2006)
[18] Li, H., Tan, J.: Body Sensor Network Based Context Aware QRS Detection. In: Per-
     vasive Health Conference and Workshops, pp. 1–8 (2006)
[19] Demongeot, J., Virone, G., Duchêne, F., Benchetrit, G., Hervé, T., Noury, N., Rialle,
     V.: Multi-sensors acquisition, data fusion, knowledge mining and alarm triggering in
     health smart homes for elderly people. C.R. Biologies 325, 673–682 (2002)
[20] Milenkovic, A., Otto, C., Jovanov, E.: Wireless sensor networks for personal health
     monitoring: Issues and an implementation. Computer Communications 29, 2521–
     2533 (2006)
[21] Doukas, C., Maglogiannis, I., Tragas, P., Liapis, D., Yovanof, G.: A Patient Fall
     Detection System based on Support Vector Machines. In: Proc of 4th IFIP Conference
     on Artificial Intelligence Applications & Innovations, pp. 147–156 (2007)

[22] Doukas, C., Maglogiannis, I., Kormentzas, G.: Advanced Telemedicine Services through Context-aware Medical Networks. In: International IEEE EMBS Special Topic Conference on Information Technology Applications in Biomedicine (2006)

[23] Lakshmi Narasimhan, V., Irfan, M., Yefremov, M.: MedNet: a pervasive patient information network with decision support. In: 6th International Workshop on Enterprise Networking and Computing in Healthcare Industry, pp. 96–101 (2004)

[24] Mihailidis, A., Carmichael, B., Boger, J.: The Use of Computer Vision in an Intelligent Environment to Support Aging-in-Place, Safety, and Independence in the Home. IEEE Transactions On Information Technology In Biomedicine 8(3), 238–247 (2004)

[25] Choudhri, A., Kagal, L., Joshi, A., Finin, T., Yesha, Y.: PatientService: electronic patient record redaction and delivery in pervasive environments. In: 5th International Workshop on Enterprise Networking and Computing in Healthcare Industry, pp. 41–47 (2003)

[26] Varshney, U.: Pervasive Healthcare. IEEE Computer Magazine 36(12), 138–140 (2003)

[27] Birnbaum, J.: Pervasive information systems. Communications of the ACM 40(2), 40–41 (1997)

[28] Khedo, K.K.: Context-Aware Systems for Mobile and Ubiquitous Networks, International Conference on Networking. In: International Conference on Systems and International Conference on Mobile Communications and Learning Technologies, p. 123 (2006)

[29] Fox, J., Beveridge, M., Glasspool, D.: Understanding intelligent agents: analysis and synthesis. AI Communications 16(3), 139–152 (2003)

[30] Zhai, J.-H., Zhang, S.-F., Wang, X.-Z.: An Overview of Pattern Classification Methodologies. In: Proceedings of the Fifth International Conference on Machine Learning and Cybernetics, pp. 3222–3227 (2006)

[31] Babic, A.: Knowledge Discovery for Advanced Clinical data Management and Analysis. In: Kokol, P., et al. (eds.) Medical Informatics Europe'99, Ljubljana, IOS Press, Amsterdam (1999)

[32] Abidi, S.S.R., Hoe, K.M., Goh, A.: Analyzing Data Clusters: A Rough Sets Approach to Extract Cluster-Defining Symbolic Rules. In: Hoffmann, F., Adams, N., Fisher, D., Guimarães, G., Hand, D.J. (eds.) IDA 2001. LNCS, vol. 2189, p. 248. Springer, Heidelberg (2001)

[33] Menachemi, N., Perkins, R.M., van Durme, D.J., Brooks, R.G.: Examining the Adoption of Electronic Health Records and Personal Digital Assistants by Family Physicias in Florida. Informatics In Primary Care 14(1), 8 (2006)

[34] Lærum, H., Karlsen, T.H., Faxvaag, A.: Effects of Scanning and Eliminating Paper-based Medical Records on Hospital Physicians' Clinical Work Practice. Journal of the American Medical Informatics Association 10, 588–595 (2003)

[35] Wang, S., Middleton, B., Prosser, L.A., Bardon, C.G., Spurr, C.D., Carchidi, P.J., Kittler, A.F., Goldszer, R.C., Fairchild, D.G., Sussman, A.J., Kuperman, G.J., Bates, D.: A cost-benefit analysis of electronic medical records in primary care. Am. J. Med. 114(5), 397–403 (2003)

[36] Hollingsworth, D.: Workflow Management Coalition, The Workflow Reference Model, TC00-1003 (Jan. 1995)

[37] HL7 Standard, http://www.hl7.org

[38] Aniruddha, G., Bharat, K., Arnaud, S.: Reinventing the Wheel? CORBA vs. Web Services, http://www2002.org//CDROM/alternate/395 (visited 11/11/2007)

[39] Dreiseitl, S., Ohno-Machado, L., Kittler, H., Vinterbo, S., Billhardt, H., Binder, M.: A Comparison of Machine Learning Methods for the Diagnosis of Pigmented Skin Lesions. Journal of Biomedical Informatics 34, 28–36 (2001)

[40] Muoz, M.A., Rodriguez, M., Favela, J., Martinez-Garcia, A.I., Gonzalez, V.M.: Context aware mobile communication in hospitals. IEEE Computer Magazine 36, 60–67 (2003)

[41] Bardram, J.: Applications of context-aware computing in hospital work: examples and design principles. In: Proc. of the ACM symposium on Applied Computing, pp. 1574–1579 (2004)

[42] Moran, T., Dourish, P.: Introduction to This Special Issue on Context-Aware Computing. Human-Computer Interaction 16(2-4), 87–95 (2001)

[43] Broens, T., van Halteren, A., van Sinderen, M., Wac, K.: Towards an application framework for context-aware m-health applications. In: Proc. of the 11th Open European Summer School (EUNICE 2005), Madrid, Spain, July 6-8 (2005)

[44] Hilario, M., Kalousis, A., Muller, M., Pellegrini, C.: Machine learning approaches to lung cancer prediction from mass spectra. Proteomics 3, 1716–1719 (2003)

[45] Prados, J., Kalousis, A., Sanchez, J.C., Allard, L., Carrette, O., Hilario, M.: Mining mass spectra for diagnosis and biomarker discovery of cerebral accidents. Proteomics 4, 2320–2332 (2004)

[46] Wagner, M., Naik, D., Pothen, A., Kasukurti, S., Devineni, R., Adam, B.L., Semmes, O.J., Wright Jr., G.L.: Computational protein biomarker prediction: a case study for prostate cancer. BMC Bioinformatics 5(26) (2004)

[47] Smith, A.E., Nugent, C.D., McClean, S.I.: Evaluation of inherent performance of intelligent medical decision support systems: utilising neural networks as an example. Artificial Intelligence in Medicine 27(1), 1–27 (2003)

[48] Futschik, M.E., Sullivan, M., Reeve, A., Kasabov, N.: Prediction of clinical behaviour and treatment for cancers. OMJ Applied Bioinformatics 2(3), 53–58 (2003)

[49] Ball, G., Mian, S., Holding, F., Allibone, R.O., Lowe, J., Ali, S., Li, G., McCardle, S., Ellis, I.O., Creaser, C., Rees, R.C.: An integrated approach utilizing artificial neural networks and SELDI mass spectrometry for the classification of human tumours and rapid identification of potential biomarkers. Bioinformatics 18(3), 395–404 (2002)

[50] Lancashire, L.J., Mian, S., Ellis, I.O., Rees, R.C., Ball, G.R.: Current developments in the analysis of proteomic data: artificial neural network data mining techniques for the identification of proteomic biomarkers related to breast cancer. Current Proteomics 2(1), 15–29 (2005)

[51] Lisboa, P.J.: A review of evidence of health benefit from artificial neural networks in medical intervention. Neural Networks 15(1), 11–39 (2002)

[52] Kohonen, T.: Self-Organizing Maps, 2nd edn. Springer, Berlin (1997)

[53] Conrads, T.P., Fusaro, V.A., Ross, S., Johann, D., Rajapakse, V., Hitt, B.A., Steinberg, S.M., Kohn, E.C., Fishman, D.A., Whitely, G., Barrett, J.C., Liotta, L.A., Petricoin, E.F., Veenstra, T.D.: High-resolution serum proteomic features for ovarian cancer detection. Endocrine Related Cancer 11(2), 163–178 (2004)

[54] Johann Jr., D.J., McGuigan, M.D., Tomov, S., Fusaro, V.A., Ross, S., Conrads, T.P., Veenstra, T.D., Fishman, D.A., Whiteley, G.R., Petricoin, E.F., Liotta, L.A.: Novel approaches to visualization and data mining reveals diagnostic information in the low amplitude region of serum mass spectra from ovarian cancer patients. Disease Markers 19, 197–207 (2004)

[55] Ornstein, D., Rayford, W., Fusaro, V., Conrads, T., Ross, S., Hitt, B., Wiggins, W., Veenstra, T., Liotta, L., Petricoin, E.: Serum Proteomic Profiling Can Discriminate Prostate Cancer From Benign Prostates In Men With Total Prostate Specific Antigen Levels Between 2.5 and 15.0 NG/ML. Journal of Urology 172(4), 1302–1305 (2004)

[56] Stone, J.H., Rajapakse, V.N., Hoffman, G.S., Specks, U., Merkel, P.A., Spiera, R.F., Davis, J.C., St.Clair, E.W., McCune, J., Ross, S., Hitt, B.A., Veenstra, T.D., Conrads, T.P., Liotta, L.A., Petricoin, E.F.: A serum proteomic approach to gauging the state of remission in wegener's granulomatosis. Arthritis Rheum. 52, 902–910 (2005)

[57] Burges, C.: A tutorial on support vector machines for pattern recognition, http://www.kernel-machines.org/

[58] Christianini, N., Shawe-Taylor, J.: An introduction to support vector machines. Cambridge University Press, Cambridge (2000)

[59] Schölkopf, B.: Statistical learning and kernel methods, http://research.Microsoft.com/~bsc

[60] Statnikov, A., Aliferis, C.F., Tsamardinos, I.: Methods for Multi-Category Cancer Diagnosis from Gene Expression Data: A Comprehensive Evaluation to Inform Decision Support System Development. Medinfo 11, 813–817 (2004)

[61] Li, L., Tang, H., Wu, Z., Gong, J., Gruidl, M., Zou, J., Tockman, M., Clark, R.A.: Data mining techniques for cancer detection using serum proteomic profiling. Artificial Intelligence in Medicine 32, 71–83 (2004)

[62] Wu, B., Abbott, T., Fishman, D., McMurray, W., Mor, G., Stone, K., Ward, D., Williams, K., Zhao, H.: Comparison of statistical methods for classification of ovarian cancer using mass spectrometry data. Bioinformatics 19(13), 1636–1643 (2003)

[63] Maglogiannis, I., Pavlopoulos, S., Koutsouris, D.: An Integrated Computer Supported Acquisition, Handling and Characterization System for Pigmented Skin Lesions in Dermatological Images. IEEE Transactions on Information Technology in Biomedicine 9(1), 86–98 (2005)

[64] Maglogiannis, I., Zafiropoulos, E.: Utilizing Support Vector Machines for the Characterization of Digital Medical Images. BMC Medical Informatics and Decision Making 4(4) (2004)

[65] Trimarchi, J.R., Goodside, J., Passmore, L., Silberstein, T., Hamel, L., Gonzalez, L.: Assessing Decision Tree Models for Clinical In-Vitro Fertilization Data. Technical Report TR03-296, Dept. of Computer Science and Statistics, University of Rhode Island (2003)

[66] Niederkohr, R.D., Levin, L.A.: Management of the patient with suspected temporal arteritis a decision-analytic approach. Ophthalmology 112(5), 744–756 (2005)

[67] Ghinea, N., Van Gelder, J.M.: A probabilistic and interactive decision-analysis system for unruptured intracranial aneurysms. Neurosurgical Focus 17(5) (2004)

[68] Markey, M.K., Tourassi, G.D., Floyd, C.E.J.: Decision tree classification of proteins identified by mass spectrometry of blood serum samples from people with and without lung cancer. Proteomics 3(9), 1678–1679 (2003)

[69] Zhu, H., Yu, C.Y., Zhang, H.: Tree-based disease classification using protein data. Proteomics 3(9), 1673–1677 (2003)

[70] Podgorelec, V., Kokol, P., Stiglic, B., Rozman, I.: Decision trees: An overview and their use in medicine. Journal of Medical Systems 26(5), 445–463 (2002)

[71] Garg, A.X., Adhikari, N.K., McDonald, H., et al.: Effects of computerized clinical decision support systems on practitioner performance and patient outcomes: a systematic review. JAMA 293(10), 1223–1238 (2005)

# AI in France: History, Lessons Learnt, State of the Art and Future

Eunika Mercier-Laurent

IAE De Lyon 6, Cours Albert Thomas 69008 Lyon, France

**Abstract.** This chapter begins by a short history of AI in France since the early 1970s. It gives some examples of industrial applications developed since the 1980s. It also introduces AFIA, the French Association for AI, and describes some activities such as the main conferences and publications. The main French AI research domains and actors such as public and private laboratories are listed and some of their activities are briefly presented. A table of AI-based French software and service companies is reprinted from the AFIA Bulletin. A presentation of the main national research programs is followed by the description of two international AI labs Sony CSL (Paris) and XRCE (Grenoble). Finally some future trends and challenges are discussed.

## 1    Introduction and Historical Sketch

To the best of my knowledge Artificial Intelligence was introduced into France by Professor Jacques Pitrat in the early 1970s. At the same time Alain Colmerauer's team in Marseille invented Prolog for natural language processing. Laforia (lip6), LRI Orsay, UTC Compiègne, IRIT Toulouse, Marseille, Nancy and LIRMM Montpellier were among the first academic AI research teams. Real interest in AI began in the early 80s. Main industrial companies initiated AI activities. Groupe Bull founded a research centre for AI in 1981 to work on natural language access to data bases, object programming and expert systems. The first European Computer Industry Research Center ECRC was founded by Bull, ICL and Siemens in 1984 and was devoted to constraint programming, design of a Prolog machine and deductive data bases. CEDIAG[1] was founded in 1985 to work on AI tools development such as KOOL, KOOL 4WD, EDEN, CHARME, Open KADS and applications [Mercier 1994]. In 1991 a global approach to corporate knowledge modelling, collecting, navigation and deployment through decision support systems was initiated [Mercier 2004]. In the middle of the 80s main industrial French and international companies such as Aerospatiale, Dassault Aviation, Dassault Electronique, IBM Scientific Centre France, CGE Marcoussis, Matra, Thomson (Thales)

---

[1]  Centre d'Etudes et Développement en Intelligence Artificielle Groupe Bull

M. Bramer (Ed.): Artificial Intelligence, LNAI 5640, pp. 91–111, 2009.

France Telecom, Rhone Poulenc, Total, Renault, Peugeot, SNCF and others had their AI teams devoted to AI applications.

The Avignon conference series on 'Expert Systems and Their Applications' (1985-1994) was the first national event bringing together research and industrial AI workers.

The first French AI software companies were born, as well as AI-based service providers.

Many industrial applications were developed during this period. The software and services companies flourished. The main applications developed in CEDIAG during this period were the following:

- ALPIN Expert system for medical insurance with natural language module for automatic processing of medical reports
- NOEMIE Configuration system for Bull computers
- Diagnosis and help desk for customer support
- KRONES configuration support system and diagnosis for bottle-washing engines
- Danish Customs' decision support system for interpretation of ECC regulations. A similar system was later developed for Argentina Customs.
- ARAMIS-GM French national Guards Missions planning system (resource allocation, crisis situation management). This first hybrid system was composed of database natural language retrieval, expert system and constraint programming techniques.
- RAMSES Security of the Winter Olympic games Albertville 1992, in which experience from ARAMIS-GM development was reused.
- SACHEM, decision support system for blast furnaces, one of the largest industrial AI projects worldwide, Sollac (Groupe Arcelor)
- Knowledge acquisition from telemetric data for Formula 1 racing cars, reusing prior experience
- Computer network diagnosis
- Optimized keys designing
- Scheduling, time-tables for colleges, universities and engineering schools
- Planning and resource allocation for orange picking and optimizing juice production [Mercier 1995], [Mercier 1996].

Many industrial people who experienced success with AI techniques and tools had no time to communicate about it. To communicate was not a part of their objectives and some applications were highly strategic and confidential. There were also many failures due to undervaluation of the difficulties and applying the same tool for solving all kind of problem. From 1994 AI was no more in fashion. Many people removed AI from their business cards; some laboratories changed the name from AI to Advanced IT.

Then the Internet came, bringing a lot of data and the information overload. It created the opportunity for new kinds of service, including e-business, e-learning, e-administration and other e-ware. In the same time the Globalization phenomenon created complex problems to solve in cross-cultural environments, the need for quick access to collective knowledge and experience, for effective distance work and decision taking, for real-time translation [Mercier 2006]. Knowledge Management became a new fashion. Due to the lack of feedback from the AI applications, the first KM tools and solutions were AI-less. When in 1996 building websites using KADS conceptual models was suggested [Verbeck and Gueye 1996] for effective retrieval, nobody believed that it could really be useful.

Globalization and the mobile phone created new needs for easy, quick and effective access to coherent information and sources of knowledge from everywhere [Mercier 1997]. Internet, Globalization, KM movement and mobile life are the extraordinary opportunities and enablers for AI techniques, tools and research. I strongly believe that web 3.0 will be *AI inside*.

French applied AI is a mix of existing techniques and hybrid ones such as Knowledge Discovery from data, text and images; Semantic web using NLP, Knowledge models and knowledge discovery techniques; Multi Agent Systems including many different techniques; hybrid solutions for complex problems such as text understanding, automatic indexing and retrieval of multimedia documents, Business Intelligence, and others.

AI is applied in various domains such as Aerospace, Insurance, Medical domains, Transportation, Design, Manufacturing, Forecast, Financial, Biotechnology, Agriculture, Military and Marketing.

*AI is inside* decision support systems for technical and medical diagnosis, help desk, maintenance, scheduling, optimization, risk analysis, process control, traffic control, design, CAI, advisory systems. AI is maybe not used enough in IP, e-ware, m-ware, HMI, Simulation, Learning, Image Processing, VR, entertainment and KM [Mercier 2006].

## 2    Actors and Topics

### 2.1    AFIA

The French Association for Artificial Intelligence was founded in 1989 to drive the French-speaking AI community, and to promote and support the development of Artificial Intelligence. AFIA is a member of ECCAI (European Coordinating Committee on Artificial Intelligence) and the founding member of ASTI (Association des Sciences et Technologies de l'Information). AFIA acts on behalf of the French AI community towards our country's public authorities, and towards ECCAI. Among AFIA members there are researchers, engineers, AI software designers, consultants, teachers, artists and students. They are frequently active in several areas of AI.

AI gathers various topics such as knowledge engineering, natural language processing, machine learning, multi-agent systems, to mention only some of them. AI has links with many other fields in computer science, including statistics, data analysis, linguistics and cognitive psychology, optimisation techniques etc. Our members are not only AI practitioners, but they also know and use "non-AI" techniques in relation to their specialities.

The main topics of interest are represented on the "AFIA diamond" below.

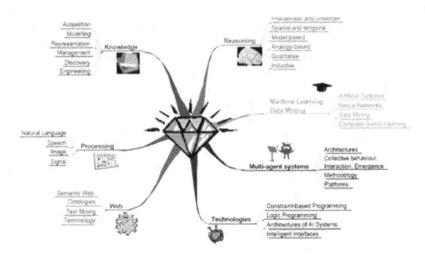

**Fig. 1.** The AFIA Diamond

The AFIA Bulletin is the publication which is the link with the members; it is also a federating tool for our community. Each Bulletin is dedicated to a specific topic as for example bioinformatics, industrial AI, Web engineering, AI and images, CBR, multimedia and AI, scheduling and others. It also contains some presentations of academic and industrial research groups focusing on specific scientific or application topics, information about conferences, books, AI publications, PhDs etc. An electronic newsletter, AFIA Infos is sent periodically to the members.

The AFIA website http://afia.lri.fr/ is designed and operated by members. Its public part contains a wealth of information, forums, links to our individual and institutional members' homepages, news of the AI community, links to conferences and other events organized and sponsored by AFIA, tutorials on AI technologies, classified announcements, etc. The private members' area gives access to a subset of the website such as studies, conference reports, book summaries, etc. It also contains a wiki for collaborative work on specific topics and allows all members to contribute to our vision on the future of AI.

### 2.1.1  Conferences and Workgroups

AFIA organised IJCAI 1993 and ECAI 2002. Since 1997 many efforts have been made to bring together several AI communities running separate conferences and industrial actors to the one common event. The first SSI[2] was held in Paris in 1998. In the next year the first Plate-forme[3] was held in Palaiseau, near Paris. It brought together several communities, such as Knowledge Engineering, Case-based Reasoning and Machine Learning. It included also ISAI[4]99 as the result of our workgroup on Operational AI applications. Since 1999 the Plate-forme has been held each odd-numbered year, the last was held in Grenoble http://afia2007.imag.fr/.

AFIA also runs colleges and workgroups. The workgroups address focused topics, can be organised as projects, and are expected to have a short or medium-term lifespan. A workgroup which grows can evolve into a "college". Our two colleges are: SMA on multi-agent systems, which itself is made up of three workgroups, and which organises annually the JFSMA conference; and CAFE (French acronym for Machine Learning, Data Mining and Knowledge Discovery), which organizes every year the CAP conference (Conference on Automated Learning).

### 2.1.2  Other AFIA Activities

AFIA also organizes Industry/Research seminars on specific topics and gives out awards and grants.

AFIA is currently the only French association whose main focus is artificial intelligence. Other associations are interested in AI and thus can have relations and share actions with AFIA. For example, collaborating with AFRIF, the French Association for Pattern Recognition, AFIA organises the RFIA "Pattern recognition and Artificial Intelligence" conference every even year. Our community is one of the first three in Europe, and is a significant contributor to European AI research.

Other French associations related to AI:

- EGC (for Knowledge Discovery and Management) http://www.polytech.univ-nantes.fr/associationEGC/

- ATALA Automatic language processing http://www.atala.org/

- PoBot (Robotics) http://www.pobot.org/

- Automates Intelligents http://www.automatesintelligents.com/

- Artificial life http://www.vieartificielle.com/

- ARCo Association for Cognitive Research http://www.arco.asso.fr/

---

2  Solutions et Systèmes Intelligents, organized by Infopromotions
3  Plate-forme is the name of the conference
4  Intelligence, Systèmes, Applications Innovantes

- Recently created consortium for AI and games http://ja.games.free.fr/Consortium_Academique/
- Consortium_academique_jeu.pdf

## 2.2    Main French Actors of AI

The Bulletin AFIA N° 49-50 published for ECAI 2002 and elaborated by Sylvie Pesty and Gilles Bisson from IMAG Grenoble included a short presentation of 141 laboratories and enterprises involved in AI research, software and applications. Some of them are listed below. The National Research System is currently being redesigned, so some of the listed labs are changing or may change in the near future.

- **CEA-DRT** (LIST-Electronics and Information Technology, LETI-Integration of Systems and Technologies, LITEN Innovation for New Energies and Nano materials) http://www-drt.cea.fr
- **CIRAD,** Montpellier, Resources & Environment, Multi Agent Systems, Collective decision making, applications
- **CRIL**, Lens, www.cril.univ-artois.fr Autonomous intelligent systems, Handling of imperfect, incomplete, context-sensitive, time-sensitive and multi-source knowledge, Inference and Decision Process
- **Crip5,** Paris, http://www.math-info.univ-paris5.fr/crip5/, Multi-Agent Systems, Knowledge Representation, Reasoning; Signal, Language and Image Processing
- **DGA**
- **DPA/DSI/AP-HP**, Paris, medical applications
- **DYNAFOR,** Castanet-Tolosan http://www.inra.fr/toulouse_dynafor. Applications of AI, Neural Networks, landscape ecology, agro-eco-systems, biodiversity
- **ENIB,** Brest http://www.lisyc.univ-brest.fr/. Five multi-disciplinary teams work on MAS for Virtual reality, bioinformatics, ecosystems modelling and simulation of complex biologic phenomena, Model engineering, process improving, learning in virtual environment, rehabilitation of handicaps in virtual environment linking VR and movement analysis, languages for automats and robots, distributed systems.
- **EAT,** Toulouse, *www.toulouse.archi.f.* Li2a team AI applied to architecture
- **EMN,** Nantes, constraint programming
- **ENSAIS,** Strasbourg, LIIA
- **ENSM-SE,** Saint-Etienne, MAS
- **ERIC**, Bron http://eric.univ-lyon2.fr
- **ERTI,** Illkirch

- **ESPCI,** Paris
- **ETIS-Imag,** Cergy
- **EURISE,** Saint Etienne, IT for environment, socio-dynamics
- **GRAVIR,** St Ismier, LAPLACE Probabilistic Models for Perception, Inference and Action)
- **GREYC,** Caen, I3 (Information, Intelligence, Interaction)
- **Grappa,** Villeneuve D'Ascq
- **HEUDIASYC – UTC,** Compiègne.
  http://www.utc.fr/recherche/Heudiasyc.php, Heuristics and Complex Systems Diagnosis: Machine Learning, Statistics, Form Recognition, Image, Decision ; Algorithms for Networks and Optimization; Documents and Knowledge; Perception and Control Systems
- **IAE,** Lyon, MODEME (Models and Methods for Advanced Information Systems
- **I2S,** Lieusaint
- **I3S,** Sophia Antipolis, TEA Techniques for Artificial Evolution
- **IFP,** Rueil Malmaison, Intelligent Information Exploration
- **IIM,** La Défense, International Institute of Multimedia
- **Innovation 3D**, St Drezery, AI for Knowledge Management and Innovation
- **INRA, MIG,** Versailles
- **INRA, UBIA,** Toulouse, Methods for Decision making, Bioinformatics, molecular biology, genetics
- **INRETS,** ESTAS, Villeneuve d'Ascq, AI applied to transport systems evaluation, automated transport and security
- **INRIA** Rocquencourt, FRACTALES (complex models and artificial evolution)
- **INRIA** Montbonnot, EXMO (Computer mediated exchange structured knowledge)
- **INRIA** Sophia Antipolis
  ACACIA (Knowledge Acquisition and Capitalization)
  AxIS (Conception, Analysis and improvement of Information Systems
  ORION (Reusable Intelligent Systems and cognitive vision).
- **INT** EPH, Evry INTERMEDIA (Interactions for Multimedia)
- **IREMIA,** Saint-Denis
  ECD (Knowledge Discovery from Data
  MAS2 (Multi Agent Systems Modelling and Simulations)
  Equipe GCC (Constraints)
- **IRIN,** Nantes, CID (Knowledge, Information and Data)

- **IRISA,** Rennes
  TEXMEX (Techniques for multimedia documents exploration)
  DREAM (Diagnostic, Recommendations and Modelling)
  CORDIAL (Multimodal Human-Machine Communication
  SYMBIOSE (Systems and models for biology, bioinformatics and sequences)

- **IRIT,** Toulouse http://www.irit.fr
  Research topics: Image Analysis and Synthesis, Indexing and Information Search, Interaction, Autonomy, Dialogue and Cooperation, Reasoning and Decision, Modelling, Algorithms and High Performance Computing, Architecture, Systems and Networks and Safety of Software Development

- **ISEN,** Lille

- **L3I** (Informatics, Image, Interaction)**,** La Rochelle
  ImagIN http://imagin.univ-lr.fr/
  Models, architectures, tools for interactive environments
  ImeDoc: Image, Media and Document http://imedoc.univ-lr.fr
  Sido: Semantic and Data Intermediation http://sido.univ-lr.fr/

- **LAAS-CNRS,** Toulouse
  DISCO (Qualitative Diagnostic, Supervision and Process Control)
  RIA (Robotics and Artificial Intelligence)

- **LAB,** Besançon
  MSF-LAB (Maintenance and Safety)
  Mobile Micro robotics

- **LAG,** Grenoble, http://www.lag.ensieg.inpg.fr
  Automatic Process Control, Safety, Supervision, Diagnostic

- **LAGIS,** University of Science and Technology, Lille, http://lagis.ec-lille.fr/
  François Cabestaing is working on EEG-based Brain-Computer Interface for Enhanced Communication [Cabestaing 2007]. Brain-Computer Interface (BCI) is a system that allows direct communication, i.e. without requiring muscular activity, between a person and a computer. In a BCI, cortical activity is recorded, analyzed and translated into orders sent to the computer. The two main approaches to EEG-based BCIs are: asynchronous, for example analyzing sensorimotor rhythms, and synchronous, for example detecting event related potentials (ERP). Further information is available from fcab@ieee.org.

- **LaLICC,** Paris

- **LaMI,** Evry, SyDRA (Distributed, Reactive and Adaptative Systems)

- **LAMIH,** Valenciennes, RAIHM (Automated Reasoning and Human-Machine Interaction

- **LAMSADE,** Paris
  SIGECAD (Knowledge and Information Systems, Decision Support)
  SMA (Multi Agents Systems)

- **LaRIA,** Amiens, IC (Knowledge Engineering)
- **LEG,** Grenoble
  CDI (Integrated Design and Diagnostic)
  Modélisation (Modeling and CAD )
- **LCIS,** Valence, CoSy (Complex Cooperative Systems
- **Leibnitz IMAG,** Grenoble, Following the new organisation of research in Grenoble in the domain of applied mathematics, computer science and signal, the Leibniz Laboratory has finished its contract. The teams are participating in new laboratories:
  o LIG, including two AI groups MAGMA (MAS) and MeTAH (Models and Technologies for Human Learning);
  o TIMC Techniques for biomedical engineering and complexity management – informatics, mathematics and applications.
    The TIMC-IMAG laboratory gathers scientists and clinicians towards the use of computer science and applied mathematics for understanding and controlling normal and pathological processes in biology and healthcare. This multi-disciplinary activity both contributes to the basic knowledge of those domains and to the development of systems for computer-assisted diagnosis and therapy.
- **LERI,** Reims MODECO-IUT (Multi Agent in Uncertain Environment Modelling)
- **LERIA,** Angers
  GII (Management of Imperfect Information)
  MOC (Metaheuristics and Combinatory Optimization)
  SBCI (Knowledge-based Interactive Systems)
  TALN&Rep de Co Natural Language Processing and Knowledge Representation
- **LESCOT,** Bron, **Ergonomics** and Cognitive Science for Transportation
- **LGP,** Tarbes, PA (Automated Production)
- **LIA,** Chambéry
- **LIA,** Avignon
- **LIF,** Marseille, BDA (Data Bases and Machine Learning
- **LIFL,** Lille, SMAC (Multi Agent and Cooperative Systems
- **LIFO,** Orléans.CA - LIFO (Constraints and Machine Learning
- **LIH,** Le Havre MAS
- **LIL,** Calais, MESC (Modelling, Evolution and Simulation of Complex Systems)
- **LIMA,** Toulouse, GRAAL (Reasoning, Action and Language)
- **LIMSI – CNRS,** Orsay
  AMI (Architectures and Models for Interaction)
  G&I (Gesture and Image)

LIR (Languages, Information and Representation)
PS (Situated Perception)
TLP (Processing of Spoken Language)

- **LIP6,** Paris

  LIP6 has a very large spectrum of research activities: networks, distributed systems, databases, languages and proofs, simulation and distributed programming, digital and symbolic computation, software for research on Computer Sciences and decision support, symbolic methods and proofs, artificial life, entity and the society of robots.

  Two main groups are involved in AI: Decision, Intelligent Systems, Operational Research and Databases and Machine Learning. The others groups are AI users.

  The first group is composed of five teams: RO (Operational research), Decision, AnimatLab, SMA (Multi agent Systems) and Mocah (Knowledge Engineering Models and Tools for Human Learning).

  o  The **Decision** team works on decision under uncertainty, multi-criteria decision making and context-based decision aiding.

  o  The **AnimatLab** is devoted to the animat approach, i.e., to the design of simulated animals or real robots, whose inner mechanisms are inspired from those of animals, and that are able to develop, learn and evolve. In the short term, the objective of the animat approach is to understand the mechanisms that make it possible for animals to adapt and survive, then to implement these mechanisms into artefacts that are also able to adapt and fulfil their mission in environments more or less changing and unpredictable. Such artefacts may be instantiated as autonomous robots that must move and explore an unknown environment, or as seemingly living characters able to interact with an human in a video game. In the long term, the objective of the animat approach is to contribute to the advancement of cognitive science by seeking in what sense human intelligence can be explained by simple adaptive behaviours inherited from animals, in a bottom-up, evolutionary and situated perspective.

In the second group, the most interesting team from the AI point of view is **ACASA**. This team, led by Jean-Gabriel Ganascia, initially focused on Knowledge Acquisition and Machine Learning. Its scientific orientations evolved to Scientific Discovery and the design of Intelligent Agents. Thus research was recently undertaken on literary analysis (genetic criticism, stylistic analysis), on modelling social representations (rebuilding of social stereotypes from newspapers), on Scientific Discovery (modelling theories in medicine), on musicology and music (detection of recurrent patterns), on multi-media, on intelligent TV and on the improvement of electronic reading facilities. Moreover, in the last few years, the ACASA team has begun work on natural language semantics, on system biology and on computational philosophy with the aim of exploring individual or social representation of knowledge, i.e. to investigate human cognition from a cultural point of view [ACASA Presentation].

- **LIPN,** Villetaneuse,
  ADAge (Machine Learning, Diagnostic and Agents)
  RCLN (Knowledge Representation and Natural Language)

- **LIRMM,** Montpellier, http://www.lirmm.fr
  Laboratory of Computer Science, Robotics, and Microelectronics
  INFO (Computer Science): constraints, learning, knowledge representation,
  multi-agent systems, data mining; Human-Computer Interaction: e-learning,
  natural language processing, hypermedia, visualization.
  Robotic Department

- **LISI,** Villeurbanne D2C (Data, Documents, Knowledge

- **LISTIC,** Le Bourget du Lac, Condillac (Ontology)

- **LIUM,** Le Mans, AI for Learning and Teaching

- **LOG,** Toulouse

- **LORIA,** Vandoeuvre-les-Nancy http://www.loria.fr/
  Natural Language Processing and multimodal communication
  Knowledge Representation and Management

- **LRI,** Orsay http://www.lri.fr
  Algorithms and Complexity
  Artificial Intelligence and Inference Systems
  Bioinformatic Team
  Inference and Learning

- **LRL,** Clermont-Ferrand

- **LSIIT,** Illkirch, AFD (Learning and Data Mining)

- **LTCI – ENST**, Paris, TII (Image Processing and Interpretation)

- **NELLE,** Colombes

- **ONERA – DCSD**, http://www.cert.fr/dcsd/en
  Systems Control and Flight Dynamics Department

- **PSI,** Mont-Saint-Aignan, DSI (Document and Interactive Systems)

- **Pôle Cindyniques,** Sophia-Antipolis,
  http://www.sophia.ensmp.fr/recherche/cindy.html
  Pôle Cindyniques, Risk Prevention and Crisis Management

- **Programme Vision HyperArtLedge**, Cesson Sévigné

- **SeT,** Belfort, MAS

- **THALES Aerospace,** Elancourt
  One of the topics this group works on is a distributed approach for coordination
  and control of several Unmanned Aerial Vehicles (UAVs) in temporally constrained missions. This approach combines multi-agent and trajectory planning
  techniques and provides a coordination model taking both deliberation and

planning durations into account. Recent advances made in the field of Unmanned Aerial Vehicles (UAVs), suggest that in the near future, fleets of UAVs will be deployed in order achieve various temporally constrained missions such as surveillance, intelligence or suppression of enemy air defences. Thus new algorithms and architectures have to be proposed to ensure a coordinated control of the fleet. This problem can be tackled in various ways, such as optimization, multi-agent simulation based on autonomous agents and using a biologically-inspired approach. Many research solutions were already proposed. The planning and deliberation process takes time and has to be handled. For this reason a distributed approach was followed, combining multi-agent and trajectory planning techniques, which in addition to allowing coordination and control of several UAVs involved in temporally constrained missions, takes both deliberation and planning durations into account. More information is available from patrick.taillibert@fr.thalesgroup.com and from [Marson, Soulignac and Taillibert, 2007].

- **Tech-CICO,** Troyes, http://tech-cico.utt.fr
  Knowledge engineering for Knowledge Management and Distributed Information Systems

- **VALORIA,** Vannes, http://www-valoria.univ-ubs.fr
  Ambient Computing, Interaction and Intelligence: providing end-users with technological, innovative means for greater user-friendliness, more efficient services support and user-empowerment, while contributing to user-friendly, dependable, adaptive and non-intrusive hardware/software environments.

AFIA Bulletins 62 and 64 (updated in July 2007) contain Gérald Petitjean's presentation of 47 enterprises working in the area of AI for decision making. They are listed in Table 1. Their activities represent the following categories:

1. **Optimization:** dynamic optimization, combinatorial optimization, constraints programming, linear programming, operational research, meta heuristics, planning, scheduling
2. **Machine Learning / Data Mining/ Knowledge Transfer:** Statistics, Data Analysis, Neural Networks, Bayesian Networks, Decision Trees, Evolutionary Algorithms, Classification, Regression
3. **Knowledge Engineering / Documents Engineering / Semantic Web / Ontologies**
4. **Multi Agents Systems**
5. **Image Processing / Vision / Forms Recognition**;
6. **Image Processing/ Natural Language Processing/ Signal Processing**
7. **Expert Systems /Logic / Reasoning** : Rules, Case-based Reasoning, Logic Programming, Fuzzy Logic
8. **Human-Machine Communication**
9. **Robotics**
10. **3D / Virtual.Reality/Games/Serious Games**

**Table 1.** Enterprises Working in AI for Decision Making

| Society | Topics | | | | | | | | | |
|---|---|---|---|---|---|---|---|---|---|---|
| | *1* | *2* | *3* | *4* | *5* | *6* | *7* | *8* | *9* | *10* |
| A2IA | | | | | X | | | | | |
| AKIO SOFTWARE | | X | | | | | X | | | |
| ALCTRA | | X | | | | | | | | |
| ALDEBARAN ROBOTICS | | | | | | | | | X | |
| APODIS | X | | | | | | | | | |
| ARDANS | | | X | | | | | | | |
| ARTELYS | X | X | | | | | | | | |
| AXLOG Ingénierie | X | | | X | | | X | | | |
| BAYESIA | | X | | | | | | | | |
| BOUYGUES E-LAB | X | | X | | | | | | | |
| CANTOCHE | | | | X | | X | | X | | X |
| CODEAS | | X | | | | | | | | |
| CO-DECISION TECHNOLOGY SAS | | X | X | X | | | | X | | |
| COSYTEC | X | | | | | | | | | |
| DAUMAS AUTHEMAN et Associés | X | | | | | | X | | | |
| EUROBIOS | X | X | | X | | | | | | |
| EURODECISION | X | | | | | | X | | | |
| EVITECH | | X | | | X | | X | X | | |
| Facing-IT | | X | | | X | | | | | |
| FircoSoft | | | | | | | X | | | |
| FRANCE TELECOM R&D Pôle Data@ledge | | X | X | | | X | X | | | |
| GOSTAI | | | | X | X | X | | | X | |
| ILOG | X | | | | | | X | | | |
| I-NOVA | | X | | | | | | | | |
| INOVIA | X | | | | | | | | | |
| INTELLITECH | | X | | | | | X | | X | |
| KAIDARA | | X | | | | | X | | | |
| KOALOG | X | | | | | | | | | |
| KXEN | | X | | | | | | | | |

Table 1 (continued)

| | | | | | | | | | | |
|---|---|---|---|---|---|---|---|---|---|---|
| **KYNOGON** | X | X | | X | | | X | | | |
| **MASA** | X | X | | X | | | | | | |
| **NORMIND** | | X | X | | | | X | | | |
| **ONTOLOGOS Corp.** | | | X | | | | | | | |
| **OSLO** | | | | X | | | | | | |
| **PACTE NOVATION** | X | X | | | | | X | | | |
| **PERTIMM** | | | X | | | X | | | | |
| **PERTINENCE** | | X | | | | | X | | | |
| **PROBAYES** | | X | | | | | | | | |
| **RENAULT DTSI/T2IA/ IAA-SICG** | X | X | | | | | X | | | |
| **ROBOSOFT** | | | | | | | | | X | |
| **ROSTUDEL** | X | | | | | | | | | |
| **SEMANTIA** | | | | | | | X | | X | |
| **SKYRECON** | | X | | | | | X | | | |
| **SOLLAN** | | | X | | | | | | | |
| **TREELOGIC** | | X | | | | X | X | X | | |
| **VECSYS** | | | | | | | X | | X | |
| **VirtuOz** | | X | | X | | X | | X | | |

## 2.3    National Research and Applied Research Programs including AI

AFIA supports the RNTL (National Network for Software Technologies) which aims at providing our country with a leading software industry building upon its excellent academic research.

AI is also included in the national programs as PRIAM and RIAM, ANR and several Pôles de Compétitivité. Pôles de competitivités were created to bring together research and industrial specialists and work on common programmes under a common development strategy. http://www.competitivite.gouv.fr

Among 71 clusters there are 7 Global competitiveness clusters such as Aerospace Valley, Finance Innovation, LYONBIOPOLE, Medicen, MINALOGIC, SCS and SYSTEM@TIC; and 10 Globally-oriented competitiveness clusters such as AXELERA, Cap Digital, Images & Networks, i-Trans, Industries & Agro-Resources, Therapeutic Innovations, MOV'EO, Pôle Mer (sea) Bretagne, Pôle Mer PACA and Végépolys. Some of them have AI teams and almost all clusters are concerned with AI as users.

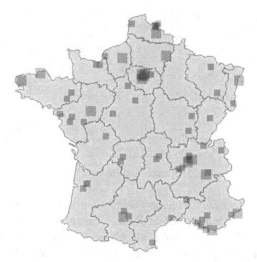

**Fig. 2.** French Poles of Competivity. Squares represent private-public research activities. The largest and the international ones are in the Paris and Grenoble areas.

**ANR** (National Agency for Research) http://www.agence-nationale-recherche.fr/ was created in January 2007 to finance innovative projects of public research and enterprise. The aim of ANR is to create new knowledge and to enable relations between the public and private research laboratories through innovative projects.

Many French AI actors are involved in European Community projects, such as IST or 7FP and others.

## 2.4 International AI Research Labs in France

### 2.4.1 SONY CSL

Sony Computer Science Laboratory Paris (Sony CSL Paris) is an offshoot of the successful Sony Computer Science Laboratory in Tokyo http://www.sonycsl.co.jp/. Both laboratories were created by Sony to perform basic research in computer science and related fields.

Sony CSL Paris research areas are language, music, sustainability, art and science, their previous work was on robotics and neuroscience.

Language is considered as a complex adaptive system that emerges through adaptive interactions between agents and continues to evolve in order to remain adapted to the needs and capabilities of the agents. The full cycle, presented in Figure 3 of speaker and hearer is explored while they play situated language games.

The 'naming game' experiments used computer simulations of communities of language users to explore the emergence of shared lexicons in a population. In the naming game, software agents interact with each other in a stylised interaction (termed a 'language game'). Repeated interactions lead to the development of a

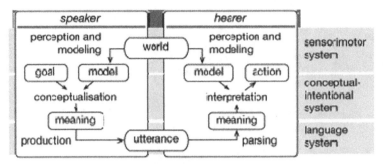

**Fig. 3.** Speaker and Hearer Cycle

common repertoire of words for naming objects. By varying experimental parameters, it is possible to explore the effect of environmental factors such as noise and uncertainty, memory limitations and contact between different language groups.

The Talking Heads experiment studied the evolution of a shared lexicon in a population of embodied software agents. The agents developed their vocabulary by observing a scene through digital cameras and communicating about what they have seen together. To add an extra level of complexity to their task, agents were able to move freely between different computer installations located in different parts of the world. Members of the public were able to influence the course of the experiment by logging on to the Talking Heads website to create and teach their own agents.

The ability to express and recognize emotions or attitudes through the modulation of the intonation of the voice is fundamental to human communication. In particular, it allows coordinating the social interactions with babies, like in language games (giving feedback, calling for attention). Any robot designed to interact with humans in a natural manner needs these skills.

Emotions have some mechanical effects on physiology, like heart rate modulation or dryness in the mouth, which in turn have effects on the intonation of the voice. This is why it is possible in principle to predict some emotional information from the prosody of a sentence.

Sony CSL are investigating how to control the pitch (fundamental frequency) and energy of synthetic speech signals so that a robot can express attitudes or emotions that can be recognized by humans. They have designed an algorithm which generates lively cartoon emotional speech, stylized so that people of many different native languages can reliably identify the emotion. The degree of emotion can continuously be varied. The technology was validated with human subjects.

Sony CSL have also made a large scale experiment in order to find out which features and which machine learning algorithms are best to recognize the emotion in the voice of human speakers. State-of-the-art data-mining techniques were used to find the best feature set among more than 400 features together with the best learning algorithm, ranging from nearest-neighbours, Bayesian classifiers, neural networks and support vector machines. Tests were made with a database of 6

Japanese speakers, with five emotions (neutral, anger, sadness, bored, happiness). There were 6000 sound samples in total. Surprisingly, it was found that the best features were not those used in the psycho-acoustic literature, but new ones based on the quartiles of the distribution of the energy values and of the minimas of the pitch contour.

Sony CSL investigates the mechanisms that enable humans and robots to learn new words and to use them in appropriate situations. They have built a number of robotic and computational experiments studying the mechanisms of concept formation, joint attention, social coordination and language games, and articulating the roles of learning, physical and environmental biases in language acquisition. The unifying theme of all these experiments is development: they explore the hypothesis that language can only be acquired through the progressive structuring of the sensorimotor and social experience.

*Sustainability*

Human economic and technological activities are beginning to impact the Earth's ecosystems at an alarming rate, leading to global warming, climate instability, stress on natural ecosystems, pollution and many other negative effects. Since 2006, CSL has begun to focus its attention very seriously on this important problem. The goal is to develop the infrastructure for *community memories* which empower a community to manage its commons with tools such as participatory sensing, environmental modelling and prediction, social tagging, geographic localisation and visualisation.

*Music*

Music research at Sony CSL Paris is concerned with three areas: the multiple facets of interaction in music, the challenges of robust music description and the mechanisms of music sharing on ad hoc networks. Dynamic relationships may exist between gesture and system response or among communities of connected users. Meta-data spans the space between high-level personal taste and low-level sonic content. Social dynamics on wireless and peer networks hold the potential to give rise to new forms of musical content. This work leads to the creation of intelligent musical systems that propose new modes of access to music, interaction with sound, and human interaction.

*Art and Science*

Art-science interactions help to stimulate intuitive thinking that is not reachable by pure rational inquiry alone. Lab members therefore regularly engage in dialogues with major artists and musicians to develop new creative works shown in major exhibition spaces and theatres.

## 2.4.2  XRCE

*Xerox Research Centre Europe http://www.xrce.xerox.com*

Xerox is a company that is founded upon and thrives on innovation. The Xerox Innovation Group explores the unknown, invents next generation technology and creates new business and shareholder value through its worldwide research centres.

Xerox established its European research centre in Grenoble, France in the early 90s to create innovative document technology and drive the corporate transition to becoming a services-led technology business. The centre coordinates research, engineering and the Technology Showroom, a showcase for Xerox research and a technology exchange forum with thousands of customers every year. XRCE also develops connections within the wider European scientific community through collaborative projects and partnerships.

XRCE is part of the global Xerox Innovation Group made up of 800 researchers and engineers in four world-renowned research and technology centres. Approximately 100 people work at the European research centre. Researchers and engineers in Grenoble collaborate with private and public research institutions in the area as well as with their Xerox colleagues at the Webster Research Centre in New York and the Palo Alto Research Center (PARC) both in the USA.

*Research & Technology*

Research in Europe takes a strongly inter-disciplinary approach, ranging from cognitive vision, computer science, statistics and mathematics to linguistics and sociology. All research projects revolve around documents and related services to improve customer communications and productivity in the workplace. There are six complementary research areas within the laboratory. Most employ machine learning and/or rule-based methodologies in multilingual text, image and data processing with the exception of the Work Practice Technology Area. This group develops a deep understanding of customer document processes to help design new technologies to support them. The applications developed at XRCE aim at streamlining document intensive processes, bridging the paper and the digital worlds and facilitating information management in multiple languages. The centre is at the heart of many of the components in Xerox's 'Smarter Document Management' suite such as text and image categorization, XML conversion and advanced linguistic analysis tools. Their deployment automates customer processes and provides value added functions such as indexing, semantic search, document routing and publishing in multiple formats.

The XRCE text categorization tool automates manual categorization and has demonstrated enormous customer cost savings and enhanced business performance. The tool works in over 20 languages.

Image categorization performs a similar function to the text categorizer but for digital images with many different kinds of content. It is useful in a number of applications in the document lifecycle including document creation using auto-illustration, digital asset management, image retrieval and image enhancement for printing.

Document conversion to the eXtensible Markup Language (XML) enables unstructured documents to be processed like structured data. The growing demand to streamline document intensive processes such as authoring and publishing or manufacturing and new drug approval is addressed by being able to automatically convert scanned documents into XML.

The underlying technology uses a unique combination of methods so that it learns and adapts itself to achieve the best results yet is reusable and hence affordable.

Text mining and semantic search tools are required too. Unstructured text is estimated to make up four fifths of the information created by companies worldwide. XRCE tools master the finely grained semantic processing required to identify business critical information for corporate finance, risk management and litigation or new drug discovery.

XRCE collaborates with academic, government and industrial research groups and participates in a variety of national and European Commission funded projects. Researchers are members of research review boards and expert panels and hold positions at other academic institutions. The centre welcomes visiting professors and students from across the world. It runs an extensive annual intern programme and also partners with universities through the Xerox Foundation which each year funds around 40 projects at 30 colleges and universities worldwide.

## 3     Future Trends

In 2006 AFIA organized a celebration of 50 years of AI[5]. The programme was elaborated to bring some interesting points of view on AI research, applications and future:

- **AI and Time for Experiences**: Jacques Pitrat

- **Back to the Future: Retrospective Glance on the Anticipations of AI**, Jean-Gabriel Ganascia (lip6)

- **Artificial Intelligence and Informatics**, Round table animated by Jérôme Euzenat (INRIA)

  o Could the information system be intelligent? Yves Caseau (Bouygues Telecom)
  o Creativity and Informatics Jean-Luc Dormoy (CEA)
  o In search of the machine to learn with, Pierre Tchounikine (LIUM)
  o Agents and Humans Yves Demazeau (IMAG)
  o Web and language : in search of relevance, Luc Steels (Sony CSL)

---

[5] The video from the 50th anniversary is on
http://ru3.com/luc/tag/afia/50-ans-intelligence-artificielle-afia.html
http://ru3.com/luc/tag/people/luc-steels-50-ans-intelligence-artificielle.html

- o  Human-machine Communication: Interaction, brain and cognitive science, François Cabestaing (Université Lille 1)

- **AI in National and European Programmes** Patrick Corsi (Kinnsys), Bertrand Braunschweig (ANR), François Cuny (Pôle System@tic),

- **AI: Towards New Frontiers, Opportunities and Challenges**
  Round table animated by Eunika Mercier-Laurent (Université Lyon 3), with the participation of Patrick Albert (Ilog), Paul Bourgine (CREA), Vincent Lemaire (France Télécom), Michèle Sebag (AFIA) and Patrick Tallibert (Thales)

Among the challenges: AI has to bring more services for today's life, help to preserve our planet, bring assistance to older people or persons with handicaps, aid us in real-time learning, creativity and problem solving [Amidon, Formica and Mercier 2005], [Mercier 2006]. The paradox is that AI have to be integrated everywhere, but when inside, AI is no more visible. To make AI visible we probably need to create a label *AI inside*.

Concerning the challenges for research – a lot of ideas can come from the "real" world - applied research is difficult but exciting.

A recent trend is to put more AI into interactive electronic games, both individual and collective, as well as into serious games. It is also to connect digital and symbolic AI as for example voice interface [Boulanger, du Chateau and Mercier-Laurent 2008].

# References

[ACASA Presentation] ACASA Lip6 Research Group presentation: Cognitive Agents and Automated Symbolic Learning, http://www-acasa.lip6.fr/index.php?choix=0&langue=en

[Amidon, Formica and Mercier 2005] Amidon, P., Formica, E., Mercier-Laurent, E.: Knowledge Economics: Emerging Principles, Practices and Policies. Tartu University Press, Tartu (2005)

[Boulanger, du Chateau and Mercier-Laurent 2008] Boulanger, D., du Chateau, S., Mercier-Laurent, E.: System for the inventory of cultural heritage using voice knowledge acquisition. Accepted for AI 2008, Cambridge (2008)

[Cabestaing 2007] Cabestaing, F.: EEG-based Brain-Computer Interface for Enhanced Communication. In: HuMaN'07 First International Conference on Human-Machine iNteraction, Timimoune, Algeria (2007)

[Marson, Soulignac and Taillibert 2007] Marson, P.E., Soulignac, M., Taillibert, P.: Combining Multi-Agent Systems and Trajectory Planning Techniques for UAV Rendezvous Problems. In: Cogis 2007, Paris (2007)

[Mercier 1994] Mercier-Laurent, E.: Right tool for the right problem. In: Expersys 94, Houston (1994)

[Mercier 1995] Mercier-Laurent, E.: Methodology for Problem Solving using AI. In: Expersys95 (1995)

[Mercier 1996] Mercier-Laurent, E.: Knowledge Management some industrial examples. In: Expersys 96, Paris (1996)

[Mercier 1997] Mercier-Laurent, E.: Knowledge based Enterprise - How to optimise the flow of knowledge - theory and practice. Knowledge Acquisition Szklarska Poreba, Poland, 25-27 April (1997)

[Mercier 2004] Mercier-Laurent, E.: From Data Programming to Intelligent Knowledge Processors. Cutter IT Journal (December 2004)

[Mercier 2006] Mercier-Laurent, E.: Brain Amplifier for Holistic Knowledge Management using New Generation of AI. In: ICAI, Beijing (2006)

[Verbeck and Gueye 1996] Gueye, P., Verbeck, F.: Open KADS & Hypertexts. Bull DSIS-CEDIAG (1996)

# Artifact-Mediated Society and Social Intelligence Design

Toyoaki Nishida

Kyoto University, Sakyo-ku, Kyoto 606-8501, Japan
nishida@i.kyoto-u.ac.jp

**Abstract.** Human society is increasingly dependent on artifacts. The progress of artificial intelligence accelerates this tendency. In spite of strong concern about heavy dependence on artifacts, it appears an inevitable consequence of the knowledge society. In this chapter, I am seeking a better way of living with advanced artifacts to realize an artifact-mediated society where people are supported by human-centered socially-adequate artifacts. The proposed framework consists of surrogates that work on behalf of the user and mediators that moderate or negotiate interactions among surrogates. I survey recent work in social intelligence design, and discuss technological challenges and opportunities in this direction.

## 1 Introduction

Human society has been dependent on artifacts since the early days of history. Conventional characterization of artifacts is to view them as tools for extending humans or resources that support us as well as artifacts.

Before the first half of the twentieth century, most artifacts extended the physical dimensions of human activities. For example, horse-drawn carriages, steam locomotives and automobiles remarkably extended our capability of moving and carrying, to name just a few. Conventional lifelines such as gas, water and power supplies, roads or railways serve as resources for enabling the activities of ourselves and artifacts. There are some others, though more limited compared with physical extensions, which extend the intellectual dimensions of human activities, such as telephones and broadcasts. The effect of classic artifacts is relatively limited to the local regions (Figure 1). Individuals are relatively isolated from each other and their human relations do not often go beyond local communities.

In the second half of the twentieth century, the invention of computers and the Internet significantly increased the complexity of artifacts. First, it brought about automation everywhere. Modern aircraft are mostly controlled and monitored with computers. ICT (Information and Communication Technologies) enabled intimate integration of mechanism, control and environment. In the case of automobiles, for example, information and communication technologies not only help drivers

M. Bramer (Ed.): Artificial Intelligence, LNAI 5640, pp. 112–132, 2009.

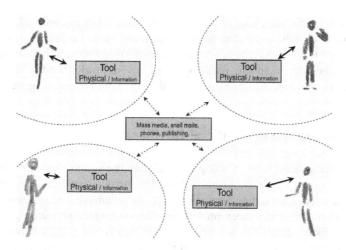

**Fig. 1.** Classic artifacts

drive an automobile safely and comfortably, but also make use of geographic positioning and information systems to refer to rich information sources concerning the driving route, parking lots, nearby restaurants or traffic conditions. Moreover, it allows the transportation authority to intelligently control traffic signals to reduce traffic jams, or place an electronic toll collection system to automatically collect a toll charge. Industrial robots are introduced to pursue long and complex assembling tasks very rapidly in factories. Second, computers and the Internet significantly extended the way we think and communicate with each other. They

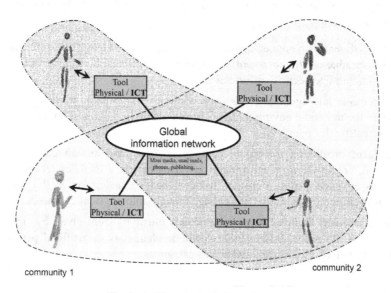

**Fig. 2.** Artifacts in the information age

provide facilities for searching as well as information. They not only allow people to talk with each other, but also help people know, meet, collaborate, help and understand each other. As a result, people come to place more value on togetherness and connectedness in life, resulting in the advent of global network communities, such as the Unix, Amazon.com, or Wikipedia communities (Figure 2).

## 2    Contribution of Artificial Intelligence

Artificial intelligence (AI) technologies accelerate the sophistication of artifacts. In its history over fifty years, AI has achieved numerous significant contributions to computer science, including heuristic search, knowledge-based systems, natural language processing and pattern processing, planning, machine learning and data mining. Heuristic search has allowed us to efficiently use partial knowledge to solve heuristically large-scale ill-formed problems, such as Chess, for which the action sequence to the solution cannot be uniquely determined in the course of problem solving due to the incompleteness of knowledge. Knowledge-based systems have enabled us to solve problems by explicitly coding experts' practice. Natural language processing and pattern processing have made computers understand what we speak, see and perceive in general. Planning enables us to implement autonomous artifacts that pursue their mission based on a set of given strategies and tactics. Machine learning permits computers to improve their behaviour based on experience. Data mining enables computers to discover patterns from a large collection of data (Figure 3).

Besides progress in the core technologies, we are witnessing increasingly more practical AI applications in numerous domains. Some remarkable applications, among others, include:

- Web intelligence by connecting a huge amount of knowledge and information made available on the net to solve challenging problems. It ranges from scientific computing to security informatics.

- Autonomous intelligent artifacts such as Mars Exploration Rovers that can explore the unknown environment of the red planet in pursuit of data collection and scientific discovery.[1]

- Intelligent prostheses for supplementing or enhancing human capability in physical, cognitive or social dimensions. This category includes technologies such as advanced safety vehicles that can locate a pedestrian in the dark or brain-machine interfaces that can change thought into physical actions.

- Artist computers that can draw, paint or play music by themselves, by simulating the artistic process. Interactive synthetic characters or artificial pets with emotion / personality models fall in this category.

---

[1]  http://marsrovers.jpl.nasa.gov/overview/

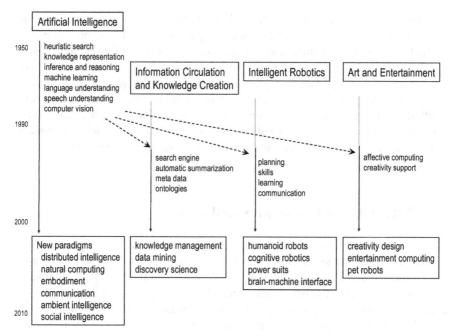

**Fig. 3.** Artificial Intelligence and adjacent areas

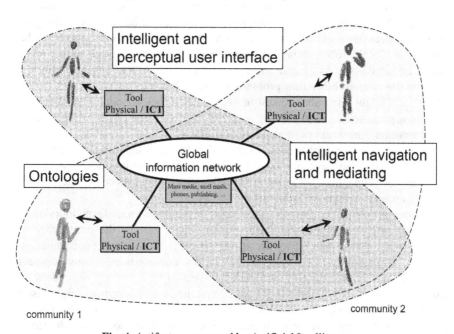

**Fig. 4.** Artifacts empowered by Artificial Intelligence

As a result, AI makes it possible to build artifacts with human-like perceptions that can execute complex social functions (Figure 4). In addition, programming methodology has significantly changed. It is becoming popular for researchers and developers quantitatively to measure phenomena, and have artifacts statistically learn from the training data to cope robustly with complex phenomena. AI will contribute to "dismantling discontinuities in historical ego smashing events", "the fourth discontinuity - the one between humans and their machine" [Cooley 2007], in particular. Although it is good news in the sense that we can implement artifacts that can work in more complex environments than before, it brings about opaque artifacts beyond our understanding.

## 3    Difficulties with Complex Artifacts

As artifacts become more intelligent, it becomes harder both for the product producers and consumers to take responsibilities for the artifacts they produce or use. On the one hand, the user cannot completely understand the artifacts so that they may not hurt or annoy other people. Indeed, the complexity of artifacts might significantly hinder their proper use. On the other hand, it is almost impossible for a producer to predict all possible uses of the product and prevent misuse from happening, even though it might be possible to manufacture error-free artifacts. The problem may become more serious with intelligent artifacts such as autonomous intelligent robots. Even though many people might believe the validity of Asimov's three laws of robotics, their complete implementation appears to be impossible, for the robot designers cannot think about all situations her/his robot would be faced with. As a result, nobody might be able to take responsibility for artifacts (the responsibility flaw problem) [Nishida 2007a].

Artifacts might bring about disastrous outcomes if they are applied to illegal or malicious purposes, or even used by a user in panic or caught by a strong antagonistic emotion. The worst example is robot soldiers.

Ethical problems might become evident, for AI technology might extend tacit human thoughts that have not caused serious problems with conventional technologies, for people may not be good at thinking tacit dimensions or society is not entirely familiar with novel problems.

Another problem is that it will cause heavy dependence on artifacts. Individuals might use artifacts without judgment. Society might assume the infallibility of artifacts without rationale. There are strong concerns about heavy dependence on artifacts. Among others, Cooley exhibited a similar concern using "from judgment to calculation" [Cooley 2007], and gave a caveat of being overly dependent on calculation rather than judgment. He warns that the lack of judgment might result in severe failures, such as a fatal dose of morphine more than 100 times the correct amount, which could have been prevented if one were trying to capture the situation with enough sensibility and deliberation. In the meanwhile, the heavy dependence will entirely remove motivations of thinking and imagination at the individual level, and might bring about "empty brains" [Maurer 2007]. Perrow points out the difficulty of sustaining sensibility against accidents as an organization, and accidents might happen normally [Perrow 1984].

Maurer warns that a serious breakdown of the computerized social infrastructure might cause a catastrophic disaster and bring human society back to the Stone Age.

Unfortunately, the heavy dependence on artifacts appears to be an inevitable destiny of mankind. The more artifacts bring about new services, the more human society may depend on artifacts, in order to overcome the complexity. We often use social heuristics to ask friends or colleagues to solve problems rather than try to solve them by ourselves. Such an attitude is often effective, for the modern world is so complex that we know explicitly or implicitly that our knowledge and information is very limited. On the other hand, the more the human society depends on artifacts, the more services may result due to the recognition of novel problems and opportunities (Figure 5).

We have already come too far deeply into "Lushai Hilles" [Cooley 2007]. Although we can see the way we passed, we cannot see what is coming up in the journey. It seems almost impossible to live without being assisted by artifacts in a modern jungle. As the above argument suggests that the difficulties come from human society rather than nature, in the rest of the chapter I will focus on issues related to assisting human society with artifacts.

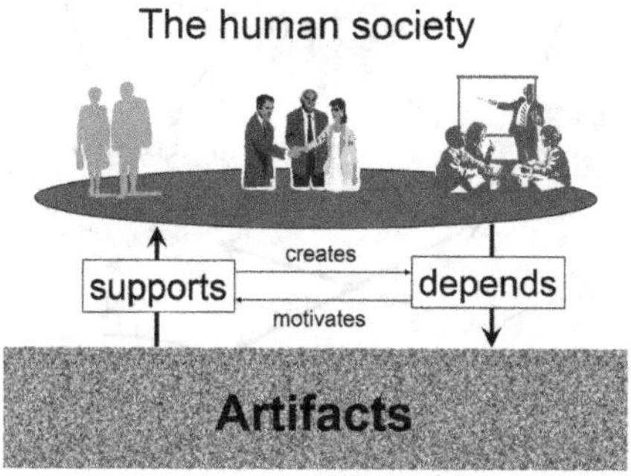

**Fig. 5.** Heavy dependence on artifacts

# 4    Towards Artifact-Mediated Society

How can we maximally draw on the power of artifacts without suffering from real and potential dangers? An intuitive answer is to design human-centered socially-adequate artifacts, or *social artifacts*, to realize an *artifact-mediated society*, in which a fully automated mechanism moderates artifacts so that all interactions among artifacts are completely moderated and malicious and inappropriate use can be suppressed or at least recorded for later assessment, even though it is not possible to completely prohibit them. I present the framework and discuss feasibility and challenges for achieving it.

## 4.1    The Framework

This framework assumes that artifacts mediate social functions. Whenever a person would like to play a social function, s/he is expected to delegate the social function to an artifact called a surrogate that may act on behalf of her/him. Mediators moderate or negotiate interactions among surrogates (Figure 6).

It is very much like people negotiating with each other through their artificial attorneys, who not only try to maximally satisfy the user's intention but also comply with legal and ethical rules shared in society. We called this approach the artifact-as-a-half-mirror approach, for a half mirror is a quasi-opaque object that can both pass light from the back and reflect the real world image. The artifacts in this paradigm may not be completely autonomous or may not be entirely amenable to the owner's intention.

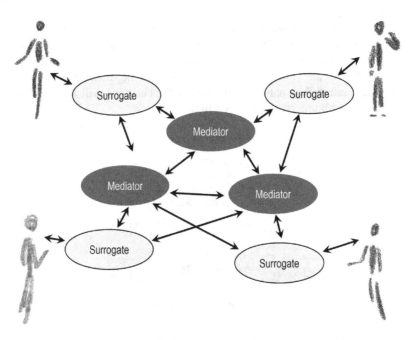

**Fig. 6.** Information explosion causes more dependence on artifacts

Figure 7 illustrates how the framework is applied to intelligent vehicles. The entire environment is moderated by an intelligent traffic controlling system consisting of mediators. Individual vehicles can be seen as one's surrogate. Unlike conventional automobiles, these intelligent vehicles will attempt to maximize the client's request, such as arriving at a specified destination in a minimal time, only if it is consistent with the public traffic rules and manners. The traffic rules may be designed to be sociable in the sense that intelligent automobiles must yield to pedestrians so long as the client is in a normal condition.

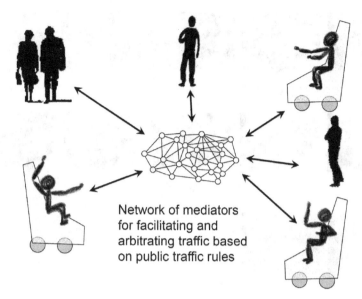

Network of mediators
for facilitating and
arbitrating traffic based
on public traffic rules

**Fig. 7.** Moderating intentions to move with autonomous vehicles

The passengers' intention such as route selection or time constraint may be passed to the socialized automobile and will be reflected in its behaviour so long as it is consistent with traffic rules.

In the autonomous mobile chair project [Terada 2002], the preliminary prototype of an autonomous mobile chair as a surrogate was implemented and tested, while no mediators were used. The autonomous chair was designed to provide a person with a place to sit down. In order to achieve its goal, the autonomous mobile chair was designed to perceive and decrease the geometric distance between its seat back and the surface of the user's body (the back). The optimal action for achieving this depends on multiple factors, such as the shape or locomotive ability of the autonomous mobile chair, or the relative angle of the two surfaces. Reinforcement learning was used to have the autonomous mobile chair learn to move to the goal position in a given situation. Although the mechanism was very simple, the automatic mobile chair worked as if it was intelligently trying to serve the user and the user's reaction toward it appeared relatively natural, as shown in Figure 8.

Although all the users hired for a small experiment were all able to sit down on the chair as a result of coordinating behaviours, some users pointed out that the autonomous mobile chair should have communicated its intentions more explicitly.

Nishida et al implemented a pair of communication robots (Figure 9) called the listener and presenter robots to elaborate the notion of robot as an embodied knowledge medium for mediating embodied knowledge among people [Nishida 2006]. The listener robot interacts with an instructor to acquire knowledge by videotaping important scenes of her/his activities. The presenter robot, equipped with a small display, will interact with a novice to show the appropriate video clip in appropriate situations where this knowledge is considered needed during her/his work.

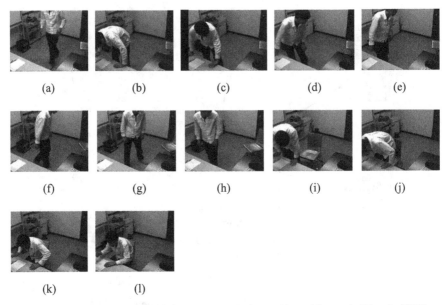

Fig. 8. Autonomous mobile chair as a surrogate interacting with people [Terada 2002]

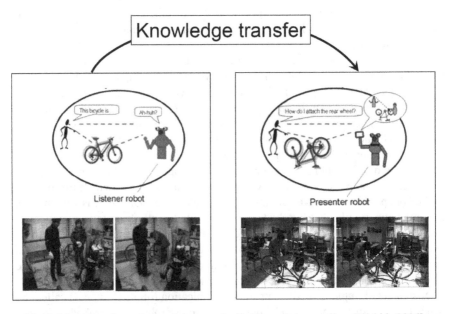

Fig. 9. Listener and presenter robots as embodied knowledge medium [Nishida 2006]

## 4.2     Challenges

There might be strong objections against using artifacts to mediate human relationships, partly because of the strong influence they might have and partly because of the potential threats of tacit flaws or brittleness to the malicious use or
abuse of the machinery. In order for the framework to be feasible and accepted by
society, several technical problems need to be solved.

### (1) Secure Social Ground

The first challenge is to establish an artificial social ground for mediators. It
should be tightly secured against malicious attacks, such as cyber terrorism.

### (2) Secure Human-Artifact Relationship

The human-artifact relationship should be properly established. Some approximation of Asimov's three laws of robotics[2] should be implemented.

### (3) Universal Usability

The second challenge is to guarantee universal usability so that everybody in society
can properly communicate her/his intention to surrogates without much difficulty.
To some degree, it is like establishing the trust relationship with a lawyer who is an
expert in legal activities. The artifact that possesses sufficient knowledge about the
problem should be able to efficiently explain it to the owner in understandable
terms. It should decrease existing divides without introducing new divides.

### (4) Ethical Computing

A theory of justice such as "each person is to have an equal right to the most extensive basic liberty compatible with a similar liberty of others" or "social and
economic inequalities are to be arranged so that they are both (a) reasonably expected to be to everybody's advantage, and (b) attached to positions and offices
open to all" [Rawls 1999] should be rigorously implemented.

### (5) Transparency for Social Inspection

Public artifacts should be transparent so that society can always inspect them for
social adequacy. The rules for programming the public mediation system should
be established. The public mediation system should be transparent for inspection.

---

[2] (1) A robot may not injure a human being, or, through in-action, allow a human being to
come to harm. (2) A robot must obey the orders given it by human beings except where
such orders would conflict with the First Law. (3) A robot must protect its own existence
as long as such protection does not conflict with the First or Second Law [Clarke 1994].

## (6) Cohabitation with Legacy System

Progress is always slow. A new innovation should be able to incorporate legacy knowledge accumulated in a conventional knowledge medium.

## (7) New Opportunities and Business Chance

People are interested in novel things. In order for new artifacts to be accepted in society, it should open up a new frontier in which new discoveries are to be made. It should increase new opportunities for job positions and business. There should be ways for improving and customizing artifacts in many ways, allowing for sophistication.

# 5    Social Intelligence Design

Issues raised in the previous section may be discussed in the context of Social Intelligence Design whose goal is understanding and augmentation of social intelligence that might be attributed to both an individual and a group. Social Intelligence Design addresses understanding and augmentation of social intelligence resulting

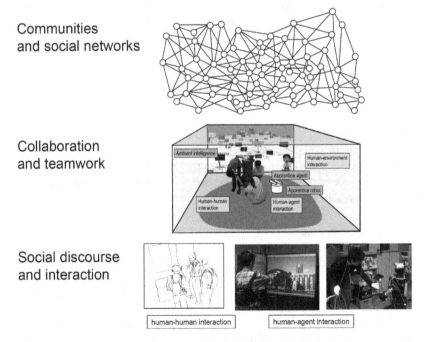

**Fig. 10.** Framework of Social Intelligence Design

from bilateral interaction of intelligence attributed to an individual to coordinate her/his behaviour with others in a society and that attributed to a collection of individuals to achieve goals as a whole and learn from experiences [Nishida 2007b]. Social Intelligence Design can be discussed at the three levels (Figure 10).

The base level comprises fast interactions at the milliseconds order where social intelligence is used to establish basic communications. The medium level encompasses collaboration in a small group to coordinate joint actions. The top level manifests at the community level to integrate individual intelligences into a collective one. The common methodology is intimate integration of analysis and synthesis based on measurement and quantitative modeling of social interactions, ranging from the small and fast to the large and slow.

## 5.1   Social Discourse and Interaction

Means of non-verbal communication such as pause, prosody, eye gaze, facial expressions, head movement, hand and body gesture, and so on constitute a basis for social discourse and interaction. Some of these behaviours are intentionally produced by a communication partner for a communicative purpose, while others, such as a subtle correlation of eye gaze and mouth movement, are not. In face to face communication, humans are considered to evaluate each other by sensing such unconscious, uncontrollable non-verbal behaviour.

Kendon and McNeill originated detailed studies on non-verbal communication [Kendon 2004; McNeill 2005]. The discourse and task level information needs to be taken into account to see the role and function of each piece of communication behaviour in the context of the entire discourse or task. Sacks introduced a concept of a turn-taking system in conversation, which suggests rules governing social interactions in conversation [Sacks 1974]. Multi-party face-to-face conversations such as those taking place at poster presentations are interesting as a subject for investigating mutual dependency between the task structure and the non-verbal communication behaviours. In the theory of participation status and framework, Goffman introduced such notions as side participants, bystanders, eavesdroppers in addition to the speaker and the addressee, in order to analyze the behaviour of actors related to a conversation [Goffman 1981].

In order for surrogates to be able to establish and maintain reliable and smooth communication with the owner, principles uncovered by detailed study should be identified and incorporated. Recent progress of measurement technologies using devices such as motion capture or eye tracker has made it possible to capture non-verbal behaviours in real time. It allows for building quantitative and computerized models for building communicative artifacts.

Conversational Informatics is a field of research for studying conversation from wider points of view encompassing conversational artifacts, conversational contents, conversational environment design, and conversation measurement, analysis, and modeling [Nishida 2007c]. A number of attempts have been made to build embodied conversational agents or communicative robots that can interact with people in a natural fashion. Kipp et al propose a data-driven approach to synthesizing natural hand and arm gestures in a continuous flow of movement [Kipp 2007]. Nakano shows that the grounding behaviours observed in human-human interaction can be effectively used to establish grounding with an embodied conversational agent [Nakano 2007]. Gratch et al reported experimental evidence that even the simple contingent non-verbal behaviours of virtual agents can contribute to establishing rapport with humans [Gratch 2007]. Kopp et al studied a framework for the feedback mechanism from a virtual agent for signaling how internal states of the listener evolve [Kopp 2007a].

Nakano and Nishida analyzed attentional behaviour in human-human interactions, and implemented the findings on embodied conversational agents. Mack can detect the user's attention to a shared referent and use attentional information to judge whether linguistic messages are grounded or not. IPOC can detect the user's attention on the display to estimate the user's interest and engagement in the conversation [Nakano 2007]. Sidner and Lee pointed out that attentional gestures such as following the looking gesture of a human partner, or displaying looking ability such as turning back to look at a human interlocutor after pointing at objects in the room contribute to sustain the sense of engagement in human-robot interaction [Sidner 2007].

André studied how face threats are mitigated by multimodal communicative acts. They conducted a user study, and revealed that gestures are used to strengthen the effect of verbal acts of politeness [Rehm 2007b]. Mohammad and Nishida studied how a social robot can express its internal state and intention to humans in a natural way using non-verbal feedback [Mohammad 2007].

Based on the "computers are social actors" paradigm, Nijholt claims that including the generation of humour and the display of appreciation of humour is useful in human-agent interaction, just as in human-human interaction [Nijholt 2007].

Kopp et al reported that there was experimental evidence for a systematic relationship between visual characteristics of the gesture form and the spatial features of the entities they refer to, though there are quite a number of false negatives. They implemented an integrated, on-the-fly planner that derives coordinated surface forms for natural language and gesture directly from communicative goals [Kopp 2007b].

From the computational point of view, fundamental principles, such as affordance, entrainment, repetition, or mutual adaptation may be needed to implement robust communicative functions. In order to increase the believability and lifelikeness of artifacts, Poel et al studied functions and effects of gaze behaviour and implemented it into a pet robot [Poel 2007]. The design of the resulting gaze behaviour system includes the gaze shifts, vergence, the vestibulo-ocular reflex and smooth pursuit. Rehm et al applied social theories, such as social impact or social influence theory to study how the user can interact with the agents to get in the agents' social network in a 3D meeting place for agents and users [Rehm 2007a].

In order to build a continuously evolving relationship with the user, a robot should be able to put itself in a mutual adaptation process (a similar phenomenon between multiple learning agents being adapting with each other). Xu et al [Xu 2007; Xu 2009] described a three stage approach consisting of a human-human WOZ experiment, a human-robot WOZ experiment, and a human-adaptive robot experiment, trying to build a prototype based on the observation of how people adapt with each other and how people improve the protocols for interacting with robots.

## 5.2    Collaboration in Small Groups

The main task of mediators is to coordinate behaviours of surrogates for collaboration. Computational aspects of collaboration have been widely studied in computer science, in particular, distributed systems and multiagent systems research. Numerous techniques of automatic resource allocation have been developed and deployed for implementing operating systems or controlling network traffic. Numerous intelligent algorithms have been developed in the research on Multi Agent Systems to negotiate agents possibly with conflicting goals, such as distributed search, problem solving and planning, distributed rational decision making including dynamic resource allocation or coalition formation, multi agent learning [Weiss 1999], and so forth.

Social communications at the higher level are for more abstract social interactions including information sharing, collaboration, negotiation, contract making, coalition, arbitration, and so on. Accordingly, the detailed study of human behaviour is vital to study high-level social interactions and their functions.

Awareness is considered to be a key for collaboration based on the viewpoint of helping people collaborate with each other by providing a context for activities. Other benefits of awareness include sharing knowledge, experiences, and feeling of connectedness. In the meanwhile, providing awareness may introduce additional costs in privacy or obligations. Markopoulos discusses the methodologies of evaluation, based on experiences with the evaluation of awareness systems [Markopoulos 2007].

Attempts have been made to observe how information media are used to help collaborative work in global teamwork or video conferences [Mark 2001; Fruchter 2001]. In addition to interaction support, information sharing is critical to enable collaboration. Recent studies involve higher order functions such as persuasion, shared understanding, ontologies, conflict negotiation and workspace design. Lievonen and Rosenberg identify space, place and setting as key notions for analyzing the impact of information and communication technologies for the design of the workspace [Rosenberg 2005].

[Stock et al 2007] propose a technology for enforcing collaboration for shifting attitudes of participants in conflict and demonstrate a collaborative tabletop interface aimed specifically for the task.

Merckel and Nishida propose using space as a reference to shared knowledge in a group [Merckel 2007]. They are developing a system called SKM (Situated Knowledge Manager) that can associate conversation quanta on varying places in the environment. An augmented reality system is being developed to retrieve the spatial coordinates in the real world from corresponding two-dimensional image points. Special emphasis is placed on minimizing the overhead for building the spatial model of the environment and operations, for if it involves extra work, it is not sustainable in a practical way.

Okamura et al [Okamura 2007] developed an augmented conversational environment for sharing in-vehicle conversations in a group. Special-purpose sensing techniques were used to capture the conversational situations in a driving simulator. For example, the system can ground the conversation on the events observed through the simulated window of the vehicle, by analyzing pointing gestures of the participants.

### 5.3    Communities and Social Networks

In order to analyze and design interactions in the large, we need viewpoints concerning human relationships such as communities or social networks [Kim 2000; Wenger 2000]. Recently, the community concept has received much attention in various contexts, such as open source communities related to the development of open source software such as Linux, SQL, or XOOPS, Web 2.0 [O'Reilly 2005], consumer communities such as Amazon.com[3] or MovieLens,[4] social networking services, or community maintained artifacts of lasting value such as Wikipedia[5]. Although some social functions at this level have been addressed, such as auction protocol [Yokoo 2005], they are rather limited. Instead, various kinds of community support systems have been proposed, developed and evaluated, most notably and deployments to support real communities.[6]

AI technologies are more concerned with higher level functions such as community display and analysis, social navigation and search, social matchmaking, event support, dispute resolution, and large scale discussion and decision making, which are being built on top of base functions such as membership management and communication infrastructure.

Community display and analysis helps the user understand her/his own community and find opportunities to contribute, as well as utilize the community asset. A classic example of community display is Referral Web which can display the human network by connecting pieces of human relations such as co-authorship [Kautz 1997].

---

[3]  http://www.amazon.com/

[4]  http://movielens.org/

[5]  http://wikipedia.org/

[6]  For example, http://www.cii.uiuc.edu/projects

Social navigation is a technique for making use of various kinds of footprints [Waxelblat 1999] of other members of a community. A classic example is social book marking [Keller 1997]. Collaborative web search is a search technique reflecting community interests and preferences by making use of previous searches of a community member with whom a common interest or preference is shared [Coyle 2007].

Social matchmaking facilitates the formation of human networks by recommending people who are considered to share common interests [Foner 1997]. Cosley proposes a method called intelligent task routing for automatically matching the human and the task in member-maintained communities [Cosley 2007].

Event support is a suite of facilities for supporting community organizers to hold regular events for community members. The ICMAS'96 Mobile Assistant Project is a classic example that encompasses communication service, Action Navigator for providing information about conference site and vicinity, and Info-Common for managing personalized information [Nishibe 1998]. Larger experiments include JSAI[7] Integrated Support System [Nishimura 2004] and CHIplace [Churchill 2004].

Dispute resolution and security are critical to maintain communities, for disputes and threats may discourage the participation of members. Mehta shows a method for detecting a profile injection attack to a collaborative filtering system [Mehta 2007].

Large scale discussion and decision making help community members exchange opinions and make decisions. Hurwitz discussed items to be considered to implement a large-scale discussion system on the net, based on the development and deployment of The Open Meeting [Hurwitz 1995]. Erickson introduced social translucence and proposed social proxy [Erickson 2009]. He implemented a system called Babble for a large scale discussion.

Sociological aspects such as workplaces or digital divide need to be considered and assessed [Rosenberg 2005; Blake 2007]. Conte and Paolucci study the effect of reputation on social networks [Conte 2007]. Caire studies conviviality as a mechanism for reinforcing social cohesion and as a tool for reducing miscoordination between people [Caire 2007].

# 6    Future Perspectives

The survey in the previous section suggests that there is still a long way to go until the agent-mediated society is realized, concerning higher-level social functions in particular. Accordingly, it might be reasonable to set as an intermediate goal the framework as shown in Figure 11, where humans make high-level decisions with information support for basic social functions.

---

[7] JSAI stands for Japanese Society for Artificial Intelligence

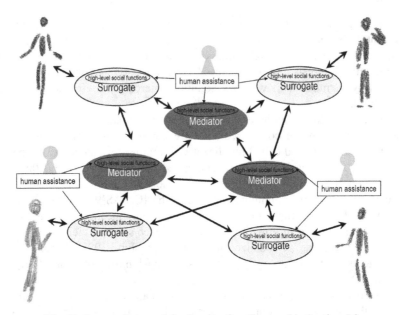

**Fig. 11.** Intermediate goal for the situation discussed in Section 5.2

Although the challenges presented in Section 4.2 can be satisfied only partially, we can gradually proceed with high-quality artifacts based on the current technologies. High-level social functions may be gradually replaced by artifacts by analyzing and incorporating the log data obtained from human-human interactions. Small-scaled social experiences will be extremely useful to obtain concrete information about how people behave in the social context, though empirical approaches based on field studies are indispensable to embed the artifacts in society.

# 7    Conclusion

In this chapter, I have discussed how Artificial Intelligence is a two-edged sword. On the one hand, it allows us to create intelligent artifacts with human-like perception and cognition. On the other hand, it accelerates people's heavy dependence on artifacts. I have discussed how to design human-centered socially-adequate artifacts. I have surveyed work related to social intelligence design and shown a future perspective.

Future work includes elaboration of ideas brought about in this chapter. In particular, drawing a 'roadmap' appears to be most effective, for it requires broad discussions across disciplines from human study and social science to engineering. Development of new technologies by integrating Artificial Intelligence with state-of-the-art measurement and interpretation techniques to bring about new insights based on human and social contexts appears to be a promising next step.

**Acknowledgment.** This chapter extends the discussions made in the International Workshop "Socio-ethics and Interactive Technologies: From Judgment to Calculation" on Thursday 4th to Saturday 6th October, 2007 at Middlesex University, Trent Park, London. The author deeply owes insights brought about by the participants, though he has responsibility on the content of this work.

# References

[Blake 2006] Blake, E., Tucker, W.: User Interfaces for Communication across the digital divide. AI & Soc. 20, 232–242 (2006)

[Caire SID 2007] Caire, P.: Designing Convivial Digital Cities: A Social Intelligence Design Approach. In: Nijholt, A., Stock, O., Nishida, T. (eds.) SID 2007 – Social Intelligence Design 2007, pp. 25–39 (2007)

[Churchill 2004] Churchill, E., Girgensohn, A., Nelson, L., Lee, A.: Blending Digital and Physical Spaces for Ubiquitous Community Participation. Communication of the ACM 47(2), 39–44 (2004)

[Clarke 1994] Clarke, R.: Asimov's Laws of Robotics Implications for Information Technology. Part 1: IEEE Computer 26(12), 53–61 (1993), Part 2: IEEE Computer 27(1), 57–66 (1994), http://www.anu.edu.au/people/Roger.Clarke/SOS/Asimov.html

[Conte 2007] Conte, R., Paolucci, M.: Reputation for Innovating Social Networks. In: Nijholt, A., Stock, O., Nishida, T. (eds.) SID 2007 – Social Intelligence Design 2007, pp. 41–53 (2007)

[Cooley 2007] Cooley, M.: From Judgment to calculation. AI & Soc. 21, 395–409 (2007)

[Cosley 2007] Cosley, D., Frankowski, D., Terveen, L., Riedl, J.: SuggestBot: Using Intelligent Task Routing to Help People Find Work in Wikipedia. In: Proc. IUI-2007, pp. 32–41 (2007)

[Coyle 2007] Coyle, M., Smyth, B.: On the Community-Based Explanation of Search Results. In: Proc. IUI-2007, pp. 282–285 (2007)

[Erickson 2009] Erickson, T.: 'Social' Systems: Designing Digital Systems that Support Social Intelligence. AI & Soc. 23, 147–166 (2009)

[Foner 1997] Foner, L.N.: Yenta: A Multi-Agent, Referral Based Matchmaking System. In: Proc. of the First International Conference on Autonomous Agents (Agents '97), Marina del Rey, California (1997)

[Fruchter 2001] Fruchter, R.: Bricks & bits & interaction. In: Terano, T., Nishida, T., Namatame, A., Tsumoto, S., Ohsawa, Y., Washio, T. (eds.) JSAI-WS 2001. LNCS (LNAI), vol. 2253, p. 35. Springer, Heidelberg (2001)

[Goffman 1981] Goffman, E.: Forms of talk. University of Pennsylvania Press, Philadelphia (1981)

[Gratch 2007] Gratch, J., Wang, N., Gerten, J., Fast, E., Duffy, R.: Creating Rapport with Virtual Agents. In: Pelachaud, C., Martin, J.-C., André, E., Chollet, G., Karpouzis, K., Pelé, D. (eds.) IVA 2007. LNCS (LNAI), vol. 4722, pp. 125–138. Springer, Heidelberg (2007)

[Hurwitz 1995] Hurwitz, R., Mallery, J.: Managing Large Scale Online Discussions: Secrets of the Open Meeting. In: Ishida, T. (ed.) Community Computing and Support Systems. LNCS, vol. 1519, pp. 155–169. Springer, Heidelberg (1998)

[Kautz 1997] Kautz, H., Selman, B., Shah, M.: Referral Web: Combining Social Networks and Collaborative Filtering. Communications of the ACM 40(3), 63–65 (1997)

[Keller 1997] Keller, R.M., Wolfe, S.R., Chen, J.R., Rabinowitz, J.L., Mathe, N.: A bookmarking service for organizing and sharing URLs. Computer Networks and ISDN Systems 29, 1103–1114 (1997)

[Kendon 2004] Kendon, A.: Gesture: Visible Action as Utterance. Cambridge University Press, Cambridge (2004)

[Kim 2000] Kim, A.J.: Community Building on the Web: Secret Strategies for Successful Online Communities. Peachpit Press, Berkeley (2000)

[Kipp 2007] Kipp, M., Neff, M., Kipp, K.H., Albrecht, I.: Towards Natural Gesture Synthesis: Evaluating Gesture Units in a Data-Driven Approach to Gesture Synthesis. In: Pelachaud, C., Martin, J.-C., André, E., Chollet, G., Karpouzis, K., Pelé, D. (eds.) IVA 2007. LNCS (LNAI), vol. 4722, pp. 15–28. Springer, Heidelberg (2007)

[Kopp 2007a] Kopp, S., Stocksmeier, T., Gibbon, D.: Incremental Multimodal Feedback for Conversational Agents. In: Pelachaud, C., Martin, J.-C., André, E., Chollet, G., Karpouzis, K., Pelé, D. (eds.) IVA 2007. LNCS (LNAI), vol. 4722, pp. 139–146. Springer, Heidelberg (2007)

[Kopp 2007b] Kopp, S., Tepper, P.A., Ferriman, K., Striegnitz, K., Cassell, J.: Trading Spaces: How Humans and Humanoids use Speech and Gesture to give Directions. In: Nishida, T. (ed.) Conversational Informatics: an Engineering Approach, John Wiley & Sons, Chichester (in press)

[Mark 2001] Mark, G., DeFlorio, P.: HDTV: a challenge to traditional video conferences? Publish-only paper, SID-2001 (2001)

[Markopoulos 2007] Markopoulos, P.: Awareness Systems: Design and Research Issues. In: Nijholt, A., Stock, O., Nishida, T. (eds.) SID 2007 – Social Intelligence Design 2007, pp. 3–12 (2007)

[Maurer 2007] Maurer, H.: Some ideas on ICT as it influences the future. NEC Technology Forum, Tokyo (2007)

[McNeill 2005] McNeill, D.: Gesture and Thought. The University of Chicago Press, Chicago (2005)

[Mehta 2007] Mehta, B., Hofmann, T., Fankhauser, P.: Lies and Propaganda: Detecting Spam Users in Collaborative Filtering. In: Proc. IUI-2007, pp. 14–21 (2007)

[Merckel 2007] Merckel, L., Nishida, T.: Solution of the Perspective-Three-Point Problem. In: Okuno, H.G., Ali, M. (eds.) IEA/AIE 2007. LNCS (LNAI), vol. 4570, pp. 324–333. Springer, Heidelberg (2007)

[Mohammad 2007] Mohammad, Y., Nishida, T.: TalkBack: Feedback From a Miniature Robot. To be presented at Australian AI Conference (2007)

[Nakano 2007] Nakano, Y.I., Nishida, T.: Attentional Behaviors as Non-verbal Communicative Signals in Situated Interactions with Conversational Agents. In: Nishida, T. (ed.) Conversational Informatics: an Engineering Approach, John Wiley & Sons, Chichester (2007)

[Nijholt 2007] Nijholt, A.: Conversational Agents and the Construction of Humorous Acts. In: Nishida, T. (ed.) Conversational Informatics: an Engineering Approach, John Wiley & Sons, Chichester (2007)

[Nishibe 1998] Nishibe, Y., Waki, H., Morihara, I., Hattori, F., Ishida, T., Nishimura, T., Yamaki, H., Komura, T., Itoh, N., Gotoh, T., Nishida, T., Takeda, H., Sawada, A., Maeda, H., Kajihara, M., Adachi, H.: Mobile digital assistants for community support. AI Magazine 19(2), 31–49 (1998)

[Nishimura 2004] Nishimura, T., Hamasaki, M., Matsuo, Y., Ohmukai, I., Nishimura, T., Takeda, H.: JSAI2003 Integrated Support System (in Japanese). Journal of JSAI 12(1), 35–42 (2004)
[Nishida 2006] Nishida, T., Terada, K., Tajima, T., Hatakeyama, M., Ogasawara, Y., Sumi, Y., Yong, X., Mohammad, Y.F.O., Tarasenko, K., Ohya, T., Hiramatsu, T.: Towards Robots as an Embodied Knowledge Medium, Invited Paper, Special Section on Human Communication II. IEICE Transactions on Information and Systems E89-D(6), 1768–1780 (2006)
[Nishida 2007a] Nishida, T., Nishida, R.: Socializing artifacts as a half mirror of the mind. AI & Soc. 21, 549–566 (2007)
[Nishida 2007b] Nishida, T.: Social Intelligence Design and Human Computing. In: Huang, T.S., Nijholt, A., Pantic, M., Pentland, A. (eds.) ICMI/IJCAI Workshops 2007. LNCS (LNAI), vol. 4451, pp. 190–214. Springer, Heidelberg (2007)
[Nishida 2007c] Nishida, T. (ed.): Conversational Informatics: an Engineering Approach. John Wiley & Sons, Chichester (2007)
[Okamura 2007] Okamura, G., Kubota, H., Sumi, Y., Nishida, T., Tsukahara, H., Iwasaki, H.: Quantization and Reuse of Driving Conversations (in Japanese). Journal of IPSJ (to appear)
[O'Reilly 2005] O'Reilly, T.: What Is Web 2.0 – Design Patterns and Business Models for the Next Generation of Software, http://www.oreillynet.com/pub/a/oreilly/tim/news/2005/09/30/what-is-web-20.html
[Perrow 1984] Perrow, C.: Normal Accidents – Living with high-risk technologies. Princeton University Press, Princeton (1984)
[Poel 2007] Poel, M., Heylen, D., Nijholt, A., Meulemans, M.: Gaze Behavior, Believability, Likability and the iCat. In: Nijholt, A., Stock, O., Nishida, T. (eds.) SID 2007 – Social Intelligence Design 2007, pp. 109–124 (2007)
[Rawls 1999] Rawls, J.: A Theory of Justice. Oxford University Press, Oxford (1999)
[Rehm 2007a] Rehm, M., Endrass, B., Wissner, M.: Integrating the User in the Social Group Dynamics of Agents. In: Nijholt, A., Stock, O., Nishida, T. (eds.) SID 2007 – Social Intelligence Design 2007, pp. 125–139 (2007)
[Rehm 2007b] Rehm, M., André, E.: More than just a Friendly Phrase: Multimodal Aspects of Polite Behavior in Agents. In: Nishida, T. (ed.) Conversational Informatics: an Engineering Approach, John Wiley & Sons, Chichester (2007)
[Rosenberg 2005] Rosenberg, D., Foley, S., Lievonen, M., Kammas, S., Crisp, M.J.: Interaction spaces in computer-mediated communication. AI & Society 19, 22–33 (2005)
[Sacks 1974] Sacks, H., Schegloff, E., Jefferson, G.: A simplest systematics for the organization of turn-taking for conversation. Language 50(4), 695–737 (1974)
[Sidner 2007] Sidner, C.L., Lee, C.: Attentional Gestures in Dialogues between People and Robots. In: Nishida, T. (ed.) Conversational Informatics: an Engineering Approach, John Wiley & Sons, Chichester (2007)
[Stock 2007] Stock, O., Goren-Bar, D., Zancanaro, M.: A Collaborative Table for Narration Negotiation and Reconciliation in a Conflict. In: Nijholt, A., Stock, O., Nishida, T. (eds.) SID 2007 – Social Intelligence Design 2007, pp. 157–165 (2007)
[Terada 2002] Terada, K., Nishida, T.: Active Artifacts: for New Embodiment Relation between Human and Artifacts. In: Proceedings of The 7th International Conference on Intelligent Autonomous Systems (IAS-7), Marina del Rey, California, USA, March 25-27 (2002)

[Waxelblat 1999] Wexelblat, A., Maes, P.: Footprints: History-Rich Tools for Information Foraging. In: CHI'99 Proceedings, ACM Press, New York (1999)

[Wenger 2000] Wenger, E.C., Snyder, W.M.: Communities of Practice: The Organizational Frontier. Harvard Business Review 78(1), 139–145 (2000)

[Weiss 1999] Weiss, G. (ed.): Multiagent Systems – A modern approach to distributed Artificial Intelligence. MIT Press, Cambridge (1999)

[Xu 2007] Xu, Y., Guillemot, M., Nishida, T.: An experiment study of gesture-based human-robot interface. In: IEEE/ICME International Conference on Complex Medical Engineering-CME2007, Beijing, China, pp. 458–464 (2007)

[Xu 2009] Xu, Y., Ueda, K., Komatsu, T., Okadome, T., Hattori, T., Sumi, Y., Nishida, T.: WOZ Experiments for Understanding Mutual Adaptation. AI & Soc. 23, 201–212 (2009)

[Yokoo 2005] Yokoo, M., Sakurai, Y., Matsubara, S.: Robust Double Auction Protocol against False-name Bids. Decision Support Systems 39(2), 241–252 (2005)

# Multilingual Knowledge Management

## Daniel E. O'Leary

University of Southern California, Los Angeles, CA 90089-0441
oleary@usc.edu

**Abstract.** Although there has been substantial research in knowledge management, there has been limited work in the area of multilingual knowledge management. The purpose of this chapter is to review and summarize some of the existing and supporting literature surrounding the emerging field of multilingual knowledge management. It does that by reviewing recent applications from multiple fields and the presentation of multilingual information. The chapter uses a theory about knowledge management and also examines supporting literature in translation, collaboration, ontologies and search.

## 1 Introduction

All types of organizations are affected by a multilingual requirement, particularly in digital environments. One of the key emerging issues associated with multinational companies and e-government is multilingual knowledge management. As those organizations face the need to provide digital knowledge resources, they also face demands for presenting multilingual digital resources as broader bases of Internet users look for digital "e-solutions." Accordingly, organizations need to determine to what extent they will provide multilingual or single language knowledge resources and address issues such as how to present multilingual knowledge resources.

In the early days of the Internet, virtually all of the search engines and content were in English (e.g., Peters and Sheriden 2000). However, in many settings that has now changed as users from all over the world, using many languages, are using the Internet. It would be too costly for firms to ignore large populations of different language speaking customers and vendors. Similarly, governments must address citizens and interested parties with different cultural and language backgrounds. As a result, global companies and governments at all levels that must provide solutions to a wider range of users are focused on providing multilingual capabilities and supporting multilingual knowledge management.

### 1.1 Purpose of This Chapter

Although the trend toward supporting multilingual corporate and government requirements is undeniable, unfortunately, there is only limited literature provid-

M. Bramer (Ed.): Artificial Intelligence, LNAI 5640, pp. 133–156, 2009.

ing an analysis of multilingual knowledge management system capabilities and applications. Most of the previous work on knowledge management has ignored multilingual issues. As a result, the purpose of this chapter is to review what is emerging as a literature of multilingual knowledge management. Although to-date there has been limited research into "multilingual knowledge management," per se, we also will examine some of the foundations and applications that are coalescing into an emerging field. More mature areas such as machine translation, are worthy of studies of their own and thus are out of the scope of this chapter.

Even the notion of what is "multilingual," appears to have multiple interpretations. In the literature, multilingual can refer to anything that involves a "multilingual user interface", to "multilingual content" or to both. Issues such as "how to do multilingual search," or "what is an ontology in a multilingual environment" need to be assessed and require further research. As a result, this chapter attempts to structure multilingual knowledge management.

### 1.2    Plan of This Chapter

Section 1 has provided an introduction and motivation for the chapter. Section 2 briefly reviews knowledge management, what it means for a system to be multilingual and some of the costs and benefits of multilingual knowledge management. Section 3 summarizes some of the research and applications of multilingual knowledge management. Section 4 discusses issues in the presentation of multilingual resources. Section 5 summarizes some of the limitations of machine translation. Section 6 analyzes some recent research from multilingual collaboration. Section 7 investigates categorization and definition of knowledge using multilingual ontologies and vocabularies. Section 8 analyzes some issues in multilingual search. Section 9 reviews some of the emerging research issues, while section 10 summarizes the chapter.

## 2    Knowledge Management and Multilingual Costs and Benefits

This section briefly reviews some basic notions of what it means to be multilingual, knowledge management, and what are some of the costs and benefits of multilingual systems.

### 2.1    Multilingual

What it means to be "multilingual" appears to occur along a spectrum. As noted by Rozic-Hristovski et al. (2002), at one extreme, multilingual means being able to select a web portal interface language. From that perspective, multilingual is almost reduced to a presentation issue. At the other end of the spectrum, not only the interface but also the resources are available in multiple languages and the links to those resources are multilingual. For example, for Peters and Sheridan (2000, p.52) multilingual refers to "…accessing, querying and retrieving information from collections in any language …." In this latter case, multilingual generally refers to content and information about the content, and connecting the user with specific aspects of the content.

As part of the user interface, multilingual also can refer to the language used to do general communication with the user of a knowledge management system, as part of multilingual presentation. As an example, a "customer survey" was made available in multiple languages (Figure 1) for a United Nations agency FAO (Food and Agricultural Organization).

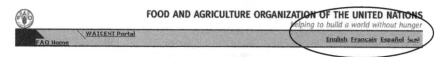

**FOOD AND AGRICULTURE ORGANIZATION OF THE UNITED NATIONS**

*elping to build a world without hunger*

WAICENT Portal

FAO Home                                            English Francais Español العربية

We are evaluating how effectively WAICENT improves access to agricultural and food security information to its users.

Please take this survey and enter the draw for exclusive prizes:

    1. 50$-gift certificate to spend on FAO publications

    2. FAO clock

    3. FAO sweater

The questionnaire consists of a minimum of 13 and a maximum of 19 questions. It should take only 10 to 15 minutes to answer.

Start the FAO Web site user survey

Comments? Please send a message to FAO-Website-Survey@fao.org                    ©FAO 2004

**Fig. 1.** Multilingual Customer Survey

## 2.2   Knowledge Management

One approach to facilitating categorization of knowledge management capabilities was presented in O'Leary (1998a) and extended by O'Leary (2008). Knowledge management systems have three primary capabilities. While providing appropriate knowledge "content," a knowledge management system may need to "convert" content to other languages and "connect" users to other users and knowledge resources (see Figure 2).

"Content" includes a broad range of resources, such as knowledge about how to solve particular problems or information about particular products or other general information capabilities. Other critical content can include ontologies used to structure and search knowledge (O'Leary 1998b), to facilitate communication and to connect knowledge. As a result, along a dimension of (*multilingual*) *content*, organizations can have the capability to provide single language knowledge resources ("the official language is English") or they can provide content in multiple languages. In between those two extremes, firms can build multilingual presentation interfaces to single language or multilingual content.

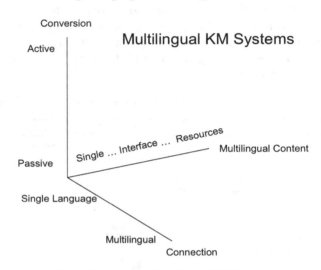

**Fig. 2.** Categorizing Multilingual KM Systems

Individual knowledge needs to be converted to group available knowledge, and data and text need to be converted to usable knowledge, not to be lost in the piles and piles of data and text that are available. In the case of multilingual systems, organizations can be located along the spectrum of the *conversion* of knowledge resources to multilingual knowledge resources. Organizations can be completely passive or actively "convert" knowledge resources to multiple languages. In a multilingual system, knowledge resources can be converted from one language to others, e.g., using collaboration and translation devices.

Further, individuals need to be "connected" to knowledge resources and other people. Knowledge resources need to be searchable and links between appropriate knowledge islands need to be established. Further, connecting knowledge to other knowledge also must consider language, since in general, knowledge in different languages cannot be consumed by all users. *Connections* between people and other people, people and resources or resources and other resources, for example search or established links, or people and other people, can be based on a single language or use multiple languages. However, ultimately, search needs to provide useful connections in the language(s) appropriate for the user.

As an example, a firm that maintains a web presence in English and Spanish, by providing the same knowledge resources in both languages, and does that by actively generating the multilingual content themselves will be at the end of all three spectra.

Historically, researchers (O'Leary 1998a) have not considered the multilingual aspects of those capabilities. As a result, from a *content, connection* and *conversion* perspective, some characterizations of knowledge management have been historically underspecified by not considering multilingual implications.

In a contemporary multilingual knowledge management system, content, conversion, and connection typically are accomplished using presentation, translation, collaboration, categorization/definition and search, among other activities. These activities and capabilities are summarized in Figure 3, along with their most frequent interactions.

| | Content | Converting | Connecting |
|---|---|---|---|
| Presentation | | | X |
| Translation | | X | X |
| Collaboration | X | X | X |
| Categorization /Definition | X | | X |
| Search | | | X |

Fig. 3. Knowledge Management Capabilities and Functions

Knowledge management content must be *presented* in a format accessible enough so that users are aware of, and can access those multilingual capabilities. For example, in a multilingual environment, users need to be able to readily change presentation language for the content from one language to another based on their native language. In this way, the user is "connected" to the "content" through the presentation. Knowledge management systems may convert content in a number of different ways. However, with multilingual systems, *translation* provides one of the key knowledge sources, using content in one language to convert to content in another language.

Collaboration can be a source of content in any knowledge management system. Further, collaboration can help convert knowledge resources from one language to another. Finally, collaboration facilitates can connect, for example, one person to another. Categorization/Definition is most likely to be accomplished in contemporary knowledge management systems using ontologies. Ontologies have emerged as important tools in multilingual knowledge management systems providing structure and definition to knowledge content that also can be used for search. However, it is also likely that those ontologies need to be connected, so that changes in one language ontology, result in changes in other language versions, and so the ontology changes seamlessly as the user moves from one language to another. Search provides a major function of connecting users to information, and capturing and indexing related information.

## 2.3    Costs and Benefits of Multilingual Systems

Multilingual capabilities are not cost free, and there are a number of potential costs and benefits of multilingual systems. First, there are maintenance costs associated with generating all knowledge resources in multiple languages. Not only are there the costs of normally generated knowledge assets in a multilingual environment, but knowledge resources must be translated to meet the needs of other users. Second, if the languages are not translated correctly or completely there can be even larger costs. For example, imagine if laws were incorrectly translated, and that inconsistent translation caused people to act in a particular way, while relying on the inconsistent translation. Third, with a multilingual system, generally, translating from one language to another takes time. If there is emergency information, many constituencies can suffer if critical information is not posted until it is available in multiple languages. Consider for example, information about a potential bird flu pandemic. If information was held up while it was translated into another language, such an action could result in substantial human destruction. Thus, time wasted by not publishing already translated information can be high. Fourth, however, if knowledge resources are translated into one language but not another, then users may perceive a bias for those constituents of the first language over those of the second. In the course of politics or with consumer groups that alienation could be quite costly. Accordingly, when to make resources available in a multilingual environment is not clear. Fifth, costs of presenting knowledge resources in one language may be less costly than other languages because of the availability and quality of translation capabilities. As a result, multilingual capabilities are subject to resource constraints and considerations. Sixth, movement to multiple languages increases the complexity of the knowledge management system. If a single language is added and that language is kept completely separate, there will be twice the resources, etc. But now imagine the interaction between each of the knowledge resources. The number of potential links between knowledge resources can explode, increasing complexity.

However, multilingual systems potentially have a number of benefits. First, by putting information into a single language, the number of users of the web material are immediately limited, no matter what the language is. Thus having resources available in multiple languages provides access to a greater base of users for important issues (e.g. Figure 4). Second, "transparency" can be increased by providing resources in more than one language. Providing materials in other languages opens up the web pages to many other potential users. Third, along with transparency, multilingual capabilities potentially generate a greater trust of the organization. Rather than hiding behind any one language a multilingual appearance provides greater access to knowledge about the organization. Finally, multilingual capabilities are likely to show a user-centric view that attempts to provide the appropriate information to the public. When a user sees an organization that provides access and content in multiple languages, it provides a view that the organization is "concerned" about the user.

**Fig. 4.** Multilingual Situation Update

# 3    Multilingual Knowledge Management Applications

In many ways, practice and the development of applications are leading theory in terms of addressing multilingual knowledge management. Organizations need to address multilingual issues now and cannot wait for theory to be developed. In particular, multilingual knowledge management is receiving attention in practice in a number of settings, including multinational firms, e-business, e-government, libraries, medicine and other international organizations.

## 3.1    Knowledge Management at a Multinational Firm

In a multinational firm, it probably is inevitable that there will be a demand for knowledge resources in multiple languages. One case study (O'Leary 2007a, p. 1142) of the large professional services firm KPMG found that the firm was concerned that different cultures, business cultures, and different languages stood in the way of the firm being a global firm. As a result, that firm ultimately kept "corporate" knowledge resources in a single language (English). However, that same firm apparently also allowed local offices to put additional servers onto the knowledge management network, and information on those servers could be placed in the originating and native languages.

Enterprises can take the position that they want a single voice / language between their employees. However, they are still likely to want to make resources, such as proposals to customers, etc. available in the native languages of their customers. Resources need to be available in those native languages so that those resources can be messaged by workers for their customers that use those native languages. As a result, ultimately multiple languages need to be accommodated, even in a firm where there supposedly is a single language.

## 3.2  E-Business Knowledge Resources

Although individual firms may be able to declare use of a single language internally, their external face generally would need to account for each language used by major groups of customers. Another study (O'Leary 2007b) investigated the web presence of twenty-five of the largest firms in the world, the so-called "Fortune 25". That research found that roughly one-half of the firms had multilingual external web presences, although only one of the twenty-five firms had more than two languages. O'Leary found that the existence of a multilingual presence apparently was generated by a number of factors. First, the dominant language was English, with only one of the Fortune 25 not providing resources in that language. Second, there was a "headquarters" effect. If there was a second language, it was likely to be the language of the country in which the headquarters of the firm was located. Third, there appeared to be an "interested parties" effect. For example, recently there has been substantial legislation in the United States aimed at regulation of enterprises. As a result, some firms disclose registration information, but only in English in order to meet the needs of the government. Fourth, there appears to be an "industry" effect. Firms in the same industry tended to disclose information in the same language as the others in the same industry. Fifth, only one organization wrapped language with culture, with the label "Hispanic". As a result, multilingual continues to refer primarily to language, and not the cultures that they typically bring with them.

## 3.3  E-Government Knowledge Resources

E-government is the provision of knowledge resources and the capability to perform governmental processes on-line, typically in a web-based digital environment. There are a number of reasons why those e-government knowledge resources also need to be multilingual. For example, a country may have multiple official languages, e.g. in Belgium French, Dutch and German are official languages. Further, a government may be part of a larger community. For example, countries in the European Union also have their own governments. As part of that larger community, countries may provide information in alternative languages beyond their official languages to facilitate "transparency."

What knowledge resources are provided by the governments in multilingual environments varies by context and government. For example, O'Leary (2007c) studied the multilingual disclosures of the United States government. In addition

**Table 1.** Summary of "Federal Citizen Information Center" Multilingual Resources by Subject
http://www.pueblo.gsa.gov/multilanguage/multilang.htm?urlnet99

| | Businesses | Civil Rights /Laws | Employment | Family, Health Safety | Money/ Benefits | News | Visitors to US | Total |
|---|---|---|---|---|---|---|---|---|
| Arabic | 1 | 1 | 1 | 1 | | | 1 | 5 |
| Armenian | | | | | 1 | 1 | | 2 |
| Cambod'n | | 1 | 1 | 1 | | | | 3 |
| Cantonese | | | | | | | 1 | 1 |
| Chinese | 1 | 1 | 1 | 1 | 1 | | 1 | 6 |
| Dutch | | | | | | | 1 | 1 |
| Farsi | | | 1 | | 1 | | | 2 |
| French | 1 | 1 | 1 | 1 | 1 | 1 | 1 | 7 |
| German | | | | 1 | | | 1 | 2 |
| Greek | | | | | 1 | 1 | | 2 |
| Haitian | | 1 | 1 | 1 | 1 | 1 | | 5 |
| Hebrew | | | | 1 | | | 1 | 2 |
| Hindi | | | 1 | 1 | | 1 | | 3 |
| Hmong | | | 1 | 1 | | | 1 | 3 |
| Italian | | | 1 | 1 | 1 | | | 3 |
| Japanese | 1 | | | 1 | | | 1 | 3 |
| Korean | 1 | 1 | 1 | 1 | 1 | 1 | 1 | 8 |
| Laotian | 1 | | | 1 | | | 1 | 3 |
| Polish | | | 1 | 1 | 1 | | 1 | 4 |
| Portug'se | | | 1 | 1 | 1 | 1 | 1 | 6 |
| Punjabi | | | | | | | 1 | 1 |
| Russian | 1 | 1 | 1 | 1 | 1 | 1 | 1 | 7 |
| Samoan | | | | 1 | | | | 1 |
| Tagalog | | 1 | 1 | 1 | 1 | | 1 | 5 |
| Thai | | | | 1 | | | | 1 |
| Ukrainian | | | | | | | 1 | 1 |
| Vietnam | | 1 | 1 | 1 | 1 | 1 | 1 | 6 |
| Totals | 7 | 9 | 15 | 20 | 13 | 9 | 18 | |

to a "Spanish" option offered on many of its web pages, information about a broad range of activities was found to be disclosed in multiple languages, as summarized in Table 1. "Family, Health and Safety" and "Visitors to the US" were the most frequently provided information. As a result, it appears as though there is a "Category of Disclosure" effect with significantly more disclosures in different categories. In addition, there was a "country effect," with differential multilingual resources provided in Korean, Russian and French.

Governments have also been concerned with other types of multilingual activities. Some other emerging areas of government multilingual interest are terrorism and crime analysis. Terrorists use multiple languages, so to find them, systems need to consider and understand multiple languages. For example, Last et al. (2006) and Qin et al. (2006) investigated the use of multilingual approaches to discover the presence of terrorist groups on the Internet. Similarly, crime does not limit itself to a single language. Thus, Yang and Li (2007) discuss how to extract multilingual information for crime analysis focusing on Chinese and English documents.

### 3.4 Library Systems

Although they were not called knowledge management systems, perhaps the first real knowledge management systems were library systems. In particular, since knowledge management systems provide access to multiple resources, one comparable source is the library, although classic library-based research is seldom directly couched as knowledge management per se.

Extending commercial notions, including personalized portals such as "My Yahoo!", there has been a sequence of research from libraries that has been related to the development of personalized library portals. Starting with Morgan (1999) and Cohen et al. (2000) libraries have allowed users to create personal web pages to capture and store frequently used electronic library resources. There have been a number of updates to that original concept and views of the future (e.g., Ciccone 2005). In addition, there have been multilingual views of the "My Library" concept. For example, as noted by Rozic-Hristovski et al. (2002, p. 157), "One of the most important needs of visitors from ... abroad is multilingual support, which means that the users can select a language in which the portal interface is presented to them."

### 3.5 Medical Systems

Sevinc (2005) and others have stressed the need for medical research to be available in multiple languages. Further, there have been some multilingual systems developed for support of medical problems. For example, Goble et al. (1994) created a multilingual terminology server designed to provide an ontology to a broad range of medical applications. As another example, Zhou, Qin and Chen (2006) focused on facilitating the search for Chinese medical information.

## 3.6     International Organizations

Some organizations are by their very nature "international." Those organizations also need to provide a range of multilingual knowledge resources. O'Leary (2008) provides an in-depth case study of one such organization associated with the United Nations, Food and Agriculture Organization (FAO). That research provides a detailed analysis of many of the multilingual issues in a large international organization. Further, that analysis examines architecture and work flow issues associated with the implementation of multilingual systems in an extensible markup language (XML) structure. That research provides a benchmark and some detailed examples about multilingual systems.

# 4     Presentation of Multilingual Resources: Connect

In general, presentation has a connection function, connecting users with knowledge resources. There are a number of presentation issues associated with multilingual resources.

## 4.1     Languages and Content Availability

Although multilingual systems may provide access to resources in multiple languages, not all resources are necessarily provided in each language to the same extent. A summary of the key issues of this section is provided in Figure 5.

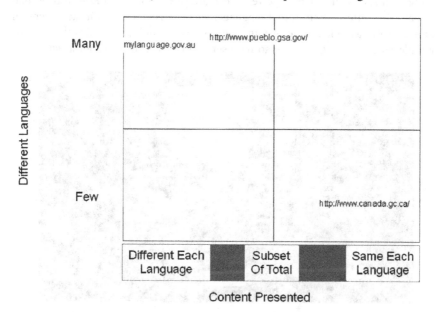

**Fig. 5.** Presentation of Multilingual Resources

At some sites all of the knowledge resources are presented for each of the languages that are accommodated. For example, at the web site for the Canadian government (http://www.canada.gc.ca/) it appears that all of the knowledge resources are provided equally in both English and French, the two official languages of the country.

At other web sites only a subset of the total knowledge resources are provided for any one language (e.g. http://www.pueblo.gsa.gov/multilanguage/multilang. htm?urlnet99), as seen in Table 1. As another example, at the Belgium web site (http://www.belgium.be/eportal/index.jsp) resources are given in four different languages (French, Dutch, German and English). However, not all resources are equally available for all languages. As a result, they provide the ability of a user to choose a "back-up" language. As noted on that site, "For the moment only a limited amount of content is available in English. In order to browse all the content of the federal portal, we suggest you to choose one of the other languages available below. Once you've done this, only the unavailable content in English will permanently be displayed in the language you have selected."

At still other sites the multilingual content in the different languages may vary substantially. For example, at mylanguage.gov.au, there is very limited overlap between the content available for the many different languages on the site.

## 4.2    Languages and Connection Links

A number of approaches have been used to capture multilingual links. As seen in Figure 1, links to other language resources are often listed on the web page that presents only information in that language, as "English," etc., with the exception of an indicator as to language. Another approach seen in Figure 6 puts the links of multiple languages side by side so the user can chose which is appropriate. This

**Fig. 6.** Slovenia Web Page (http://www.gov.si/)

approach provides a notion that the resources are transparent, and equitable, in that neither language is provided with greater knowledge resources.

Another approach is to provide links to foreign language content in another language. For example, in the following Figure 7, links are to French content (http://www.pueblo.gsa.gov/multilanguage/multilang.htm#French). However, some of the links are in English and some are in French. It is not clear if English links to French materials are an effective means of presenting the material, and will generate use of foreign language content. Further, such issues also must be considered for search, which includes links and content.

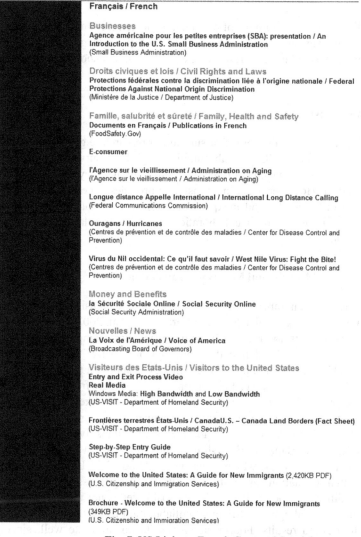

**Fig. 7.** US Links to French Content Materials

### 4.3     What Makes a Page?

Presentation often is thought of as translation. Translating different languages can result in documents of different lengths. Presentation then needs to assess if the different lengths are appropriate. Further, cultural issues need to be brought to the presentation of information to ensure that the resulting translation and presentation is appropriate for the setting.

## 5     Translation: Convert and Connect

Translation will help meet capabilities of conversion and connection. For example, resources in one language are translated to those of another language, and users in one language are connected to those resources, e.g., through collaboration or other approaches.

### 5.1     Machine Translation

Machine translation has been the source of substantial research (e.g. Nirenburg et al. 1994 and others). As a result, it is substantial enough for a survey of its own, without other knowledge management topics. Accordingly, it is beyond the scope of the current chapter. However, machine translation provides hope that resources in one language can be rapidly and inexpensively converted to other languages. Further, through activities such as collaboration, machine translation can help connect people to people.

### 5.2     Limitations of Multilingual Machine Translation

Unfortunately, there are a number of limitations associated with using translations for multilingual systems (e.g., Inaba 2007 and Caracciolo et al. 2007). First, trans-lating different languages is not necessarily "symmetric". For example, the French word "banque" can be translated into the English word "bank." However, the English word "bank" can refer to a place where money is kept or a place besides a river. As a result, in the latter case, the appropriate translation from English to French would be "rive." Such errors and ambiguities are more likely to be intro-duced into translated text the more languages into which a language is translated. Second, translating different languages is not necessarily "transitive." For exam-ple, using Google, "River Bank" in English translates to "Rive" in French, which translates to "Ufer" in German, which translates to "Shore" in English. Third, there are differential translation capabilities for different languages. The extent of translation capabilities and the quality of those capabilities is not uniform across all language pairs. Accordingly, multilingual collaboration using translation may not result in the desired results. Fourth, acronyms do not translate well, since words in different languages are likely to start with different letters and/or occur in a different order. As a result, unless the acronyms become accepted words in multiple languages, they are difficult to translate. Fifth, translations of official names as opposed to unofficial and shortened names also must be accounted for (e.g., United States, vs. United States of America).

# 6     Multilingual Collaboration: Content, Connecting and Converting

One approach to connecting users to users or users to content, and converting what they know to explicit knowledge is collaboration. Increasingly, collaboration attempts to facilitate multilingual capabilities. However, at this time, multilingual collaboration is primarily occurring in research labs.

Multilingual collaboration occurs in a number of forms. In its easiest to implement multilingual form, multilingual users contribute in multilingual settings. Unfortunately, the number of multilingual users is limited and the number of languages that any one individual communicates in is limited. As a result, in some cases users can employ their own language and that language will be translated.

## 6.1     Users Contribute in Multiple Languages

Perhaps the easiest from a system perspective is to facilitate collaboration where the users are multilingual. In this setting, users might be able to go from one conversation to another collaboration to another collaboration, independent of translation capabilities.

## 6.2     Users use their own language and it is translated

Unfortunately, generally, we cannot count on the user to have multilingual capabilities. Even if they have multilingual capabilities, they are likely to have access to a limited number of languages, and that access is likely to be stronger in some languages than others.

As a result, there has been substantial research examining how to facilitate multilingual collaboration through translation. Nomura et al. (2003) tested communication and collaboration in five different languages on a multilingual bulletin board system. They found that machine translations were problematic, impairing communication. As a result, they allowed multilingual users to modify translated sentences to improve the overall level of the translation. Funakoshi et al. (2003) developed a tool with which they were able to experiment with multilingual collaboration. They found that although translation may be appropriate for overall and high level discussions, that it was not appropriate for "detailed" discussions of cooperative works. Inaba et al. (2007) proposed a "language grid" structure to support multilingual content, with an active human user community that, like in Nomura et al (2003) allowed human participation in addition to the machine translation.

### 6.3    Human – Machine Integration

The limitations of machine translation led Nomura et al. (2003), Inaba et al. (2007) and others to describe a number of collaboration tools for multilingual environments. Throughout, their use of tools integrates human and machine capabilities to facilitate a better understanding of the necessary translation in an effort to improve multilingual collaboration.

# 7    Categorization and Definition: Content and Connecting

Knowledge categorization and definition help facilitate content and help connect content to users. The primary contemporary approach to categorizing and defining knowledge is to use ontologies.

Ontologies and semantic devices, such as controlled vocabularies also facilitate multilingual knowledge management. In particular, each can be used to facilitate multilingual content and multilingual search that can connect users to content and other users. Further, each can be used to help standardize content across multiple languages.

### 7.1    Ontologies

Ontologies are used to specify content and facilitate search. As a result, they are critical to multilingual environments. Gruber (1993) referred to ontologies as an explicit specification of a conceptualization. Gruber (1993) also suggested "contexts" to capture local views of a domain, as opposed to the global view of an ontology.

Kahng and McLeod (1996) realized that building a shared ontology would never be easy. As a result, they suggested creating ontologies with static and dynamic aspects. The static components would be mutually understood between participants, while the dynamic aspect evolves either by adding ontologies or discovering them. Using this approach, more than a single name can be given to the same concept.

Segev and Gal (2007) based their work on Kahng and McLeod (1996) and Gruber (1993) when they proposed an ontology-based model of multilingual applications, with static components and dynamic components. The static components would be that portion mutually agreed upon, while dynamic would relate to a particular language. Their model was based on a global ontology that was manually designed for a specific domain. In addition, their model used local contexts, to further specify the ontology. The combination of ontologies and contexts lends itself to multilingual applications, where a single global ontology fails to capture all of the nuances that stem from language and cultural differences.

## 7.2    Example: WordNets

There are a number of multilingual concept-based dictionaries available world wide. Fensel (2004) briefly discusses a multilingual version of EuroWordNet

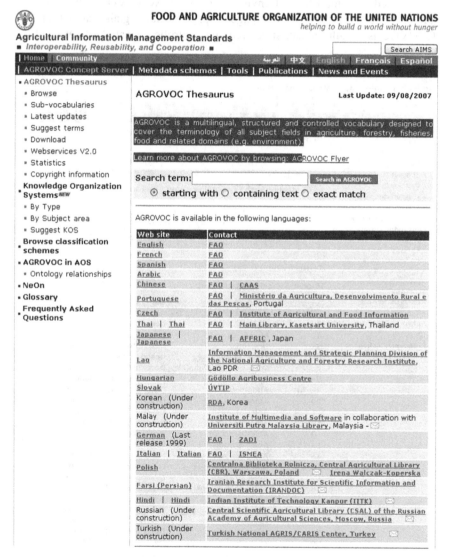

**Fig. 8.** AROVOC

(http://www.illc.uva.nl/EuroWordNet/). The word nets are structured nets of synonymous words with semantic relations between them. Because they are linked, it is possible to go from words in one language to other similar words in another language. Among the different word nets there is a static top shared ontology.

### 7.3     Example: Controlled Vocabularies such as AGROVOC

Lauser et al. (2002) examine some of the issues associated with developing an ontology used in a multilingual environment, with AGROVOC, illustrated in Figure 8. As noted on the FAO web page (http://www.fao.org/aims/ag_intro.htm) "AGROVOC is a multilingual, structured and controlled vocabulary designed to

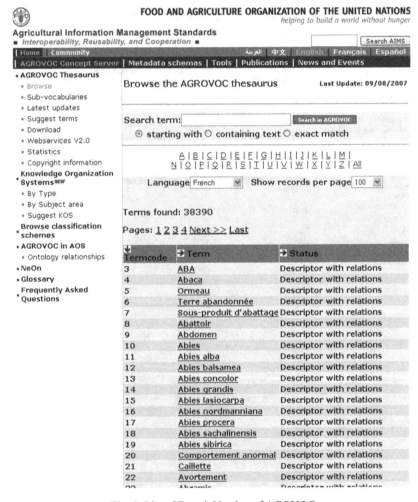

**Fig. 9.** List of French Version of AGOVOC

cover the terminology of all subject fields in agriculture, forestry, fisheries, food and related domains (e.g. environment)." As seen in the exhibit, apparently, some versions are FAO resources (e.g., the English, French, etc. versions), while others are controlled at a country-specific location, e.g., the Lao version.

According to Caracciolo et al. (2007), translations are provided by native speakers of the target language. Translations are typically made of the English version and sent to FAO for validation and inclusion in the master version. Apparently, terms are assigned a unique number, e.g. "Abalone" is assigned the number 5 in English. The translation of the word, e.g. in French "ormeau" is also given the number 5 in the French version. As a result, the vocabulary is not alphabetically ordered in each language (see Figure 9), but multiple names are attached to a single concept across the languages. AGROVOC is implemented using inheritance and a relational database structure.

The number of terms in the AGROVOC vocabulary apparently varies by language, with the number of entries for FAO controlled versions by language listed in the following Table 2. Interestingly, the number of terms in the vocabulary varies substantially by language. The differences can be the result of differences in language, but also control over additions to the vocabularies in those languages.

Information about the stability of such vocabularies is limited. Information about the impact of that stability on the use or expansion of these vocabularies is also limited. Further, based on these different sizes of the vocabularies, either some vocabularies are under-specified or some are over-specified, or characteristics of languages differentiate themselves from other language vocabularies for the same set of concepts.

**Table 2.** AGROVOC Vocabulary Size in Different Languages

| Language | November 1, 2007 |
| --- | --- |
| Arabic | 25948 |
| Chinese | 36794 |
| Czech | 39466 |
| English | 39613 |
| French | 38390 |
| Japanese | 38659 |
| Portuguese | 36347 |
| Spanish | 41714 |
| Thai | 25420 |

# 8     Multilingual Search: Connecting and Content

Search connects the user to the knowledge content. Issues include but are not limited by concerns over the number of letters in the alphabets being searched. Not surprisingly, the search literature is substantial and complete analysis is beyond the scope of this chapter. Here we will examine some specific issues that differentiate multilingual search.

## 8.1     Selected Issues in Multilingual Search

There are a number of issues associated with multilingual search. First, is the user interface of the search engine available in multiple languages? As seen in Figure 10, one search engine at FAO provides interfaces in five different languages.

Second, does the search engine allow the user to chose the language of materials that they want to search for? For example, does a search engine permit the use of an English interface, in the search for French materials about a particular topic? Third, does the search engine provide a user with a list of all the relevant materials, independent of language? For example, when using a French language interface, should the default be to search for French language materials? These and other issues can be addressed in the analysis of multilingual search.

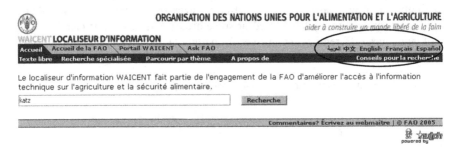

**Fig. 10.** Search in an FAO Knowledge Base

## 8.2     Non-English Search

There has been substantial research on search in general, and search in multilingual settings (e.g., Savoy 2005). For example, Bar-Ilan and Guttman (2003) analyzed the ability of three different search engines to handle queries in four non-English languages. They found that content from languages that were not English had a larger chance of being lost in cyberspace.

Non-English search is difficult for a number of reasons. As an illustration, there can be special characters in languages. For example, Aytac (2005) noted the difficulty of doing searches in Turkish because of those special characters. As another illustration, Caracciolo et al. (2007) review some other issues, including the need to use particular character encoding capabilites (UTF-8 vs. UTF-16) and the need to support left to right and right to left languages.

## 8.3    Multilingual Portals

Portals provide users with a summary of key available knowledge resource materials. From a multilingual perspective, portals can be simply a multilingual interface structure to non-multilingual content, or the portals can lead to multilingual content. An example of a multilingual portal interface is given in Figure 11.

Multilingual portals can function as a multilingual user interface, putting all of the same content in different languages. On the other hand, multilingual portals also could be constructed by ensuring that the interface and the linked content are multilingual. A decision must also be made as to whether cultural differences are sufficiently large so that the information linked to and listed on the page should be the same for each language.

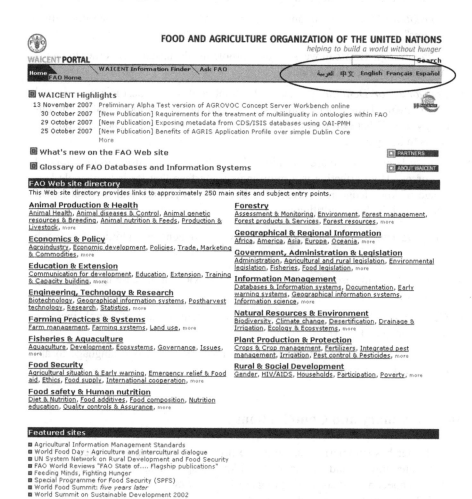

**Fig. 11.** Portal with Multilingual Capabilities

## 9    Extensions and Research Issues

Using the research summarized in this chapter there are a number of issues that require additional research. Many of the additional research questions relate directly to the topics discussed above, presentation, translation, collaboration, categorization/definition and search including the following topics.

- *Presentation:* What is the impact of delaying multilingual disclosures until all languages are ready? How effective is it to present links to foreign language material not in the foreign language? If resources are not available at the same time for multiple languages, under what conditions should they be made available to their users – should all versions be released at the same time?

- *Translation:* How efficient and effective are human-machine based translation systems?

- *Collaboration:* How can we overcome problems of translation in order to facilitate multilingual collaboration?

- *Categorization/Definition*: What kinds of ontologies work best in a multilingual environment? How much change occurs in ontologies and does that inhibit ontology use? What portion of a multilingual ontology is static and what portion needs to be dynamic?

- *Search:* How effective are multilingual portals in guiding users to multilingual resources? How effective is a multilingual interface that leads to a single language of knowledge resources? One key issue is should, and to what extent does search include multilingual content?

Further, we can use the theoretical framework to generate additional research issues, such as the following:

- *Content*: Does the use of multiple languages provide an increase in transparency and trust?

- *Connection*: Does including multilingual content as part of search increase transparency and trust?

- *Conversion*: Does converting knowledge resources from one language to another create transparency and trust?

## 10    Summary and Contributions

This chapter has summarized some of the primary multilingual knowledge management literature around a basic model of knowledge management capabilities of content, connecting and converting (e.g., O'Leary 1998a). Further, this chapter has extended some of the theory of knowledge management beyond that of early developments to account for multilingual aspects of knowledge management. In addition, this chapter investigated how those capabilities were implemented in some of the key knowledge management activities, including presentation, translation, collaboration, categorization/definition and search. Applications in a number of domains were summarized. In addition, this chapter has pointed to a number of potential research topics.

# References

Aytac, S.: Multilingual Information Retrieval on the Internet: A Case Study of Turkish Users. The International Information & Library Review 37, 275–284 (2005)

Bar-Ilan, J., Gutman, T.: How do search engines handle non English Queries? A Case Study. In: Proceedings of the Twelfth International World Wide Web Conference (2003)

Caracciolo, C., Sini, M., Keizer, J.: Requirements for the treatment of multilinguality in ontologies within FAO (October 2007)

Ciccone, K.: MyLibrary @ NCState,: A Library Portal After five Years. Journal of Library Administration 43(1/2), 19–35 (2005)

Cohen, S., Fereira, J., Horne, A., Kibbee, B., Mistlebauer, H., Smith, A.: MyLibrary: Personalized Electronic Services in the Cornell University Library. D-Lib Magazine 6(4) (2000), http://www.dlib.org/dlib/april00/mistlebauer/04mistlebauer.html

Fensel, D.: Ontologies: A Silver Bullet for Knowledge Management and Electronic Commerce, 2nd edn. Springer, Berlin (2004)

Funakoshi, K., Yamamoto, A., Nomura, S., Ishida, T.: Lessons Learned from Multilingual Collaboration in Global Virtual Teams. In: Tenth International Conference on Human Computer Interaction (2003), http://www.ai.soc.i.kyoto-u.ac.jp/publications/03/kaname-hcii2003.pdf

Goble, C., Crowther, P., Solomon, D.: A Medical Terminology Server. In: Karagiannis, D. (ed.) DEXA 1994. LNCS, vol. 856, Springer, Heidelberg (1994)

Gruber, T.: A translational approach to portable ontologies. Knowledge Acquisition 5(2), 199–220 (1993)

Inaba, R., Murakami, Y., Nadamoto, A., Ishida, T.: Multilingual Communication Support Using the Language Grid. In: Ishida, T., Fussell, S.R., Vossen, P.T.J.M. (eds.) IWIC 2007. LNCS, vol. 4568, pp. 118–132. Springer, Heidelberg (2007)

Kahng, J., McLeod, D.: Dynamic classification ontologies for discovery in cooperative federated databased. In: Proceedings of the First International Conference on Cooperative Information Systems, Brussels, Belgium, June 1996, pp. 26–35 (1996)

Last, M., Markov, A., Kandel, A.: Multi-lingual Detection of Terrorist Content on the Web. In: Chen, H., Wang, F.-Y., Yang, C.C., Zeng, D., Chau, M., Chang, K. (eds.) WISI 2006. LNCS, vol. 3917, pp. 16–30. Springer, Heidelberg (2006)

Lauser, B., Wildeman, T., Poulos, A., Fisseha, F., Keizer, J., Katz, S.: A Comprehensive Framework for Building Multilingual Domain Ontologies: Creating a Prototype Biosecurity Ontology. In: Proceedings of the International Conference on Dublin Core and Metadata for e-Communities, pp. 113–123. Firenze University Press, Firenze (2002)

Morgan, E.: MyLibrary @ NCState. In: Proceedings of the Customized Information Delivery Workshop, SIGIR, Berkeley, CA, August 19,1999, pp. 12–18 (1999), http://infomotions.com/musings/sigir-99/

Nirenburg, S., Carbonell, J., Tomita, M.: Machine Translation: A Knowledge-based Approach. Morgan Kaufman Publishers, San Francisco (1994)

Nomura, S., Ishida, T., Yosuyoka, M., Yamashita, N., Funakosh, K.: Open Source Software Development with your Mother Language: Intercultural Collaboration Experiment 2002. In: Proceedings of the International Conference on Computer Supported Cooperative Work, HCI2003, pp. 1163–1167 (2003)

O'Leary, D.E.: Knowledge Management Systems: Converting and Connecting. IEEE Intelligent Systems and their Applications 13(3), 30–33 (1998a)

O'Leary, D.E.: Using AI in Knowledge Management: Knowledge Bases and Ontologies. IEEE Intelligent Systems and their Applications 13(3), 34–39 (1998b)

O'Leary, D.E.: Enterprise Resource Planning Systems. Cambridge University Press, Cambridge (2000)

O'Leary, D.E.: Evolution of Knowledge Management Toward Enterprise Decision Support: The Case of KPMG. In: Bernsein, F., Holsapple, C. (eds.) Handbook on Decision Support Systems, pp. 1135–1162 (2007a)

O'Leary, D.E.: Multilingual E-Government. unpublished paper (Sept. 2007b)

O'Leary, D.E.: Multilingual Web Presence: Case of the Fortune 25. WeB 2007, Montreal (Dec. 2007c)

O'Leary, D.E.: A Multilingual Knowledge Management System: A Case Study of FAO and WAICENT. Decision Support Systems, 45(3), 641–661 (2008)

Peters, C., Sheridan, P.: Multilingual Information Access. In: Agosti, M., Crestani, F., Pasi, G. (eds.) ESSIR 2000. LNCS, vol. 1980, pp. 51–80. Springer, Heidelberg (2001)

Qin, J., Zhou, Y., Reid, E., Lai, G., Chen, H.: Unraveling International Terrorist Groups' Exploitation of the Web: Technical Sophistication, Media Richness, and Web Interactivity. In: Chen, H., Wang, F.-Y., Yang, C.C., Zeng, D., Chau, M., Chang, K. (eds.) WISI 2006. LNCS, vol. 3917, pp. 4–15. Springer, Heidelberg (2006)

Rozic-Hristovski, A., Humar, I., Hristovski, D.: Developing a Multilingual, Personalized Medical Library Portal: Use of MyLibrary in Slovenia. Electronic Library and Information Systems 37(3), 146–157 (2002)

Savoy, J.: Comparative Study of Monolingual and Multilingual Search Models for Use with Asian Languages. ACM Transactions on Asian Language Information Processing 4(2), 163–189 (2005)

Segev, A., Gall, A.: Enhancing Portability with Multilingual Ontology-based Knowledge Management. Decision Support Systems, Forthcoming (Available online August 6, 2007)

Sevinc, A.: Multilingual Approach to 'Web of Science. Journal of the National Medical Association 97(1), 116–117 (2005)

Yang, C.C., Wei, C.-P., Li, K.W.: Cross Lingual Thesaurus for Multilingual Knowledge Management. Decision Support Systems, Forthcoming (2007)

Yang, C., Li, K.: An Associate Constraint Network Approach to Extract Multi-lingual Information for Crime Analysis. Decision Support Systems 43(4), 1348–1361 (2007)

Zhou, Y., Qin, J., Chen, H.: CMedPort: An Integrated Approach to Facilitating Chinese Medical Information Seeking. Decision Support Systems 42(3), 1431–1448 (2006)

# Agents, Intelligence and Tools

Andrea Omicini, Michele Piunti, Alessandro Ricci, and Mirko Viroli

ALMA MATER STUDIORUM, Università di Bologna, Cesena, Italy
{andrea.omicini, michele.piunti, a.ricci,
mirko.viroli}@unibo.it

**Abstract.** This chapter investigates the relationship among agent intelligence, environment and the use of tools. To this end, we first survey, organise and relate many relevant approaches in the literature, coming from both within and without the fields of artificial intelligence and computer science. Then we introduce the A&A meta-model for multiagent systems (MAS), where *artifacts*, working as tools for agents, are used as basic building blocks for MAS modelling and engineering, and discuss the related metaphor of the *Agens Faber*, which promotes a new, principled way to conceive and build intelligent systems.

## 1 Introduction

The role of *tools* beyond language is variously exploited in human activities and societies. Organised workspaces based on artifacts and tools of diverse nature are ubiquitous in human *environments*. A tool can be conceived and explicitly built to achieve a specific goal (embedding a specific goal), stored for repeated and iterated use and exploited for building new tools. In general, a tool requires expertise of its users, their awareness of the domain problem, as well as their expertise in problem solving [14]. In all, the ability to use and make tools is as essential as symbolic language skills in defining intelligence of human beings and is typically used by ethologists to understand and measure animal intelligence.

Given the straightforward anthropomorphic interpretation of agents as human representatives in computational systems, even in the trivial acceptance of *personal assistants* [13], the need for a definition of a notion of tool in the fields of MAS (multiagent systems) and AI is quite obvious. In particular, it seems essential for the very notion of intelligence in MAS and in general for the notion of intelligent system to provide agents with the conceptual instruments to perceive and affect the environment where they live and interact, going beyond the well-explored issues of an agent's internal architecture and ability to speak symbolic languages.

In this chapter we first survey some of the most relevant conceptual frameworks in human sciences where the notions of environment and tools are suitably developed and related to the issue of human intelligence (Section 2). Then, we shift our focus on artificial systems (Section 3), discussing some of the most relevant literature on the relationship between intelligence and environment. Drawing from the results of the previous sections, in Section 4 we present the A&A meta-model for

M. Bramer (Ed.): Artificial Intelligence, LNAI 5640, pp. 157–173, 2009.

MAS, which introduces a further dimension besides agent rationality in the context of intelligent agent systems, i.e. the dimension of the artifacts and tools conceived and designed to support agent rationality and activity. The notion of *Agens Faber* is discussed, and we elaborate on its impact on the notion of agent and system intelligence, as well as on the construction of intelligent systems. Finally, Section 5 concludes the chapter.

## 2   The Role of Environment, Artifacts and Tools in Human Cognitive Systems

The history of human evolution is characterised by development of forms of complex social life accompanied by growing cognitive abilities in using and making complex tools communicating in complex ways [7]. However, as remarked by Norman [16], the power and importance of culture and artifacts to enhance human abilities are ignored within much of contemporary cognitive science despite the heavy prominence given to their importance in the early days of psychological and anthropological investigation. The field has a sound historical basis, starting at least with Wundt [29], nurtured and developed by the Soviet social-historical school of the 1920s [26,12,11,27], and still under study by social scientists, often unified by titles such as *activity theory, action theory, situated action,* with most of the research centered in Scandinavia, Germany, and the former Soviet Union.

In the early part of the 1900s, American psychology moved from its early interest in mental functioning to the behavioral era, in which studies of representational issues, consciousness, mind, and culture were almost neglected. As a result, the historical continuity with the earlier approaches as well as with European psychology had been lost. With the end of the behavioral era, American cognitive psychology had to recreate itself, borrowing heavily from British influences. The emphasis was on the study of the psychological mechanisms responsible for memory, attention, perception, language, and thought within the single, unaided individual, studied almost entirely within the research laboratory. There was little or even no emphasis on group activities, on the overall situation in which people accomplished their normal daily activities, or on naturalistic observations, and then little thought was given to the role of the environment (whether natural or artificial) in the study of human cognition.

The field only recently returned to pay serious attention to the role of the situation, other people, natural and artificial environments, and culture. In part, this change has come about through the dedicated effort of the current researchers, in part because the current interest in disciplines such as Computer Supported Cooperative Work, Human Computer Interaction and Distributed Artificial Intelligence has forced consideration of the role of real tasks and environments, and therefore of groups of cooperating individuals, of artifacts, and of culture.

In the remainder of the section we discuss the main points that characterise the notion and role of environment, artifacts and tools within human cognitive systems, by recalling some of the main concepts developed in the context of the studies and disciplines that mostly focussed on such aspects.

## 2.1  Context and Tools in Human Activities: Activity Theory and Distributed Cognition

The basic underlying principle of Activity Theory (AT) is the *principle of unity and inseparability of consciousness (human mind) and activity*: human mind comes to exist, develops, and can only be understood within the context of a meaningful, goal-oriented, and socially determined interaction between human beings and their material environment. Then, a fundamental aspect for AT has been from its beginning the *interaction* between the individuals and the *environment* where they live, in other terms, their *context*. After an initial focus on the activity of the individuals, AT research has lately evolved toward the study of human collective work and social activities, then elaborating on issues such as the coordination and organisation of activities within human society.

A central point in the AT conceptual framework is the fundamental role of *artifacts* and *tools* in human activities: according to AT every non-trivial activity is mediated by some kind of artifact. More precisely, every activity is characterised by a *subject*, an *object* and by one or more *mediating artifacts*:

- a *subject* is an agent or group engaged in an activity;
- an *object* (in the sense of *objective*) is held by the subject and motivates the activity, giving it a specific direction (the objective of the activity); the object of activity could range from mental objectives (e.g. making a plan) to physical ones (e.g. writing a paper);
- the *mediation artifacts*, which are the tools that enable and mediate subject actions toward the object of the activity. The mediating artifacts could be either physical or abstract / cognitive, such as symbols, rules, operating procedures, heuristics, scripts, individual / collective experiences, and languages.

The definition is clearly oriented to bringing into the foreground not only individuals (subjects) and their cognitive aspects, but also the context where they play, and the continuous dynamic process that links subjects to the context.

Mediation, along with *mediated interaction*, is a central point of the definition. This reflects one of the main conceptual cornerstores coming from Soviet Psychology (SP): there, a fundamental feature of human development is the shift from a *direct* mode of acting on the world to one *mediated* by some external tool; whereas Marx focused on physical mediating tools, Vygotsky extended the concept toward psychological tools. In both cases a human actor, according to the vision, is not reacting directly to the world, as an animal does, but always by means of a mediating artifact of some sort.

According to AT, mediating tools have both an *enabling* and a *constraining* function: on the one hand, they expand out possibilities to manipulate and transform different objects, but on the other hand the object is perceived and manipulated not 'as such' but within the limitations set by the tool. Mediating artifacts shape the way human beings interact with reality. According to the principle of internalisation / externalisation, shaping external activities ultimately results in shaping internal ones. Then, artifacts embody a set of social practices, and their design reflects a history of particular use They usually reflect the experiences of other people who have tried to solve similar problems at an earlier time and invented / modified the tool to make it more efficient. Experience is accumulated in the structural properties of the tools (shape, material,..), and in the knowledge of how the tools should be used as well The term *appropriation* is used to indicate the process of learning these properties and knowledge [26]. Finally, mediating tools are created and transformed during the development of the activity itself, then they carry on a given culture, the historical remnants of that development. So, the use of tools is a means for the *accumulation* and *transmission* of *social knowledge*: tools influence not only the external behaviour, but also the mental functioning of individuals using them.

The notion of artifact and tool are also at the core of Distributed Cognition (DCog), a theory of psychology recently developed by Edwin Hutchins [9], focussing on the social aspects of cognition. The core idea of DCog is that human knowledge and cognition are not confined to the individual: instead, they are distributed by placing memories, facts, or knowledge on the objects, individuals, and tools in our environment. Accordingly, social aspects of cognition are understood and designed by putting emphasis on the environment where individuals are situated, which provides opportunities to reorganise the distributed cognitive system to make use of a different set of internal and external cognitive processes. A system is conceived as a set of representations, which could be either in the mental space of the participants, or external representations available in the environment. Here, a main aspect concerns the interchange of information between these representations. For this purpose, DCog proposes a framework where the co-ordination between individuals and artifacts is based on explicitly modelling the representations where information is held in and transformed across, as well as the processes by which representations are co-ordinated with each other.

### 2.2 Intelligence as Active Externalism

In [24], Sterenly illustrates how many organisms are epistemic and ecological engineers: their evolutionary capabilities to alter systems are part of their intelligence, and often have fitness effects that get strengthened across generations. In particular, cognitive agents that have the particular epistemic ability to intentionally act for changing the informational character of their environment are used to facilitate their activities. Humans continually modify and arrange their environment not only in order to achieve some personal goal, but also for additional practical aims. By modifying environments, they ease their tasks aiding memory and computational burdens.

Clark's *Active Externalism* [5] described the active role of environment in driving cognitive processes, blurring the boundaries between internal and external cognition. In particular, he enlightened the special relation between mind and artifacts.

Both internal and external resources are viewed as an *extended mind* engaged in a larger coupled interaction where the agent (the subject) interacts with an external device (the tool). Sometimes such boundaries are indiscernible, and the aggregate agent-plus-environment may be seen as generating a unique extended cognitive system. An example was brought by Simon [23], according to whom human internal reasoning relies on internal resources such as brain memory 'as if' they were external devices, hence search information in personal memory is not different in essence from search in the external environment.

Besides, Sterelny [24] stressed that the effective use of artifact and epistemic tools relies on informational resources internally present to the agent. He noticed that the use of epistemic tools in common and contested space, where artifacts are jointly and repeatedly modified, created and used by a society of agents, *remains* a cognitively demanding process. Agents require a rich information base to fully exploit artifacts: they acquire this information piecemeal, often with the need to integrate different properties and uses across different information domains. Moreover, the use of artifacts in society implies the need to resolve problems of social and normative coordination (negotiation, division of tasks and resources).

## 2.3 Language vs. Tool, or the Language as a Tool

Gordon H. Hewes [8] observed how only in recent years has the relation between language and tools in human evolution finally been perceived as a single coherent problem. Language is an example of how subjective, inner cognitive abilities have been externalised in the outside world. Language evolved to enable extensions of human cognitive resources in the context of social, actively coupled systems: it clearly allows humans to properly spread some of their cognitive burden to others. In doing so, language serves as a special mediating tool whose role is to extend the bounds of subjective cognition. One may envisage the strict relation between use of language as a particular instance of the more general use of tools, both expanding the boundary between organisms (or agents) and the environment where they live.

In this context, along with language, tools can be considered as a distinctive expression of intelligence, powerful amplifiers of the (both individual and social) animal ability to affect the environment either to adapt to changes, i.e. preventing risks, or to meet purposes and achieve goals. Ethologists commonly assess the boundaries meant for intelligence by observing the ability of animals in facing problems that require the use of tools to be solved (see for instance [2,21]). Even more interestingly, a *tool-equivalent* of the Turing test has been originally proposed by philosopher Ronald Endicott, and then refined by R. St. Amant and A.B. Wood in [2]. The "Tooling Test for Intelligence" is aimed at evaluating intelligence in terms of the ability to use, build and share tools - see also [2] for a detailed description. In their childhood, humans learn to manipulate objects even before evolving language abilities. Their linguistic and technical skills merge together in the processes of cognitive development. Speech, manual gesture, and other forms of communication evolve side by side to tool-using skills and imitative abilities [7]. Accordingly, artifacts and tools are objects that invite humans to develop new (non-lexical-based) interactions in their environment, hence improving skills of use, affordance, recognition, and manipulation.

# 3    The Role of Environment, Artifacts and Tools in Artificial Cognitive Systems

In the sciences of the artificial, the notion of environment has essentially followed the same conceptual trajectory as in human sciences. Neglected in the very beginning by the developments of the "Symbolic AI" traditional approach, the notion of environment was subsequently strongly developed by Brooks' situated agents - however, essentially refusing cognitive aspects of intelligent behaviour. Agre & Horswill [1] paved the way for a full-fledged notion of environment to be integral to the agent's reasoning loop, while Kirsh [10] pointed out the fundamental principles for the intelligent use of the environment.

After summarising such influential approaches to the notions of environment and tool in the remainder of this section, the next section presents the *Agens Symbolicus plus Agens Faber* vision, which extends the analytic approaches by Agre & Horswill and Kirsh towards the synthetic issues of the construction and engineering of intelligent systems, by introducing and exploiting the A&A meta-model for MAS.

## 3.1    Revisiting the Concept of Environment in AI and Robotics

The definition of the notion of environment in AI and robotics is somehow fuzzy, and essentially derives from the subjective view of agents that see the environment as whatever is "outside their skin". In general, the environment in artificial systems is defined, in a sense, dually with respect to the active entities in the system, and typically includes everything relevant (places, objects, circumstances, ...) that surrounds some given agent. The environment is where agents live, it is either the target or the means of their actions, and as such it determines their effects. In natural / social systems, as well as in artificial systems, a model of the environment is required, which could account for its structures and dynamics, and could work as a stable basis for an agent's practical reasoning.

The distinction between agent and environment has been somehow blurred in early AI: the world where agents live and interact is typically oversimplified, and can typically be reduced to the subjective perception and individual action of the agent. The notion of "task environment" by Newell & Simon [15] is on the one hand perfectly functional as a framework for analysis of action and environment, on the other hand it is also nothing more than a formal representation of the space of choices and outcomes that hides the complexity of the physical world, and also essentially reduces world to a subjective agent construct. As a result, for instance, two agents co-existing within the same environment but featuring different abilities to reason and act will basically result in inhabiting different task environments, even though their physical surroundings and goals might be identical. While this bias may bring no problems in cases such as theorem-proving and chess, it turns out to be crucial whenever agent's actions have uncertain outcomes.

Newell and Simon's phenomenological approach to task analysis was overcome by Agre and Horswill [1], who explicitly introduce a notion of environment out of the agent mind to support agent reasoning. In particular, the concept of *lifeworlds* introduces a structured vision of the environment, organised so as to promote the activities that take place within it. In particular, similar to [10], lifeworlds organise and arrange tools in the environment so as to make agent's activities simpler, and reduce the cognitive burdens of agents. Tools mediate activities and spread suitable solutions to given problems, and their use promotes repeated and customary activities. Environments may contain artifacts that have been *specifically evolved*, and thus designed, to support agent activities. Having in mind the precise dynamics of agent activities it becomes possible to design suitable machinery (such as tools, artifacts) that are consistent with a given pattern of interaction. Thus, principled characterisation of interactions between agents and their environment is then explicitly used to guide explanation and design.

## 3.2  The Intelligent Use of the Environment

To explain how the structure of environments may provide an empowerment in cognitive abilities, Kirsh adopts the metaphor of the environmental 'oracle' [10]. Accordingly, the environment can suitably be used as an external memory, for example for reminding the system which tasks still have to be performed, for spatially clustering the task in sub-problems, or for organising the tools and the artifacts according to some ordering principle. Thus, tool discovery, selection and use can be considered as a fundamental part of agent intelligence, as well as a pivotal aspect for defining agent activities. In [10], David Kirsh illustrated a number of reasons to shift attention to the particular role of environment for complex and distributed problem solving. By analysing agents and environments in their *relation of use*[1], Kirsh noticed many important benefits for agents.

**Using environment simplifies choice** — Intelligent use of space speeds up the creation of a problem space representation. Besides, this helps to contract the complexity and the cognitive costs by reducing the *fan-out* of the feasible actions and the number of times at which to take a decision. Exploiting tools ameliorates the average branching factor, hence the complexity of the decisions that was made less complex by information that could be read off from environment. It eliminates decision points, creating more rational decision alterna-

---

[1] A widely accepted definition of tool use is due to Beck [3] *"tool use is the external employment of an unattached environmental object to alter more efficiently the form, position or condition of another object, another organism, or the user itself when the user holds or carries the tool during or just prior to use and is responsible for the proper and effective orientation of the tool"*.

tives[2]. Moreover it allows better heuristics of choice, driven by the affordances of tools and their perceived utilities.

**Using environment simplifies perception** — Space arrangements have a direct influence on costs of perception, i.e. making it possible to notice properties that have not been noticed yet, facilitating the discovery of relevant information. Simple techniques *regionalizing* environments and clustering objects according to their similarities can be used to highlight their differences thus creating categories of use and restricting the kind of actions an agent may take. Exploiting landmarks or pro-actively marking objects can be a practice for reminding activities and information not to be lost.

**Using environment saves computation** — Particular setting of environments can be suitably exploited to save internal computation, for instance creating a visual cue in order to make the relevant information more explicit.[3]

## 4   Agents and Artifacts as Abstractions for Engineering Intelligent Systems

In the context of autonomous agents and multi-agent systems (MAS), the notion of environment recently gained attention for the design and engineering of intelligent systems [28]. Besides agents, as autonomous pro-active entities, either software or physical, designed to accomplish some kind of goal or perform some activity, the environment can be considered a *first-order entity* defining the context where agents are situated, designed so as to embed some functionalities and features to be exploited by agents for supporting their individual and collective work.

Among others, the Agents and Artifacts (A&A) approach [22] introduces the notion of *artifact* as a first-class abstraction for modelling and designing MAS working environments, i.e. that part of the MAS designed by MAS engineers and then *used* by agents at runtime for their goals or tasks. The conceptual and engineering framework introduced by A&A draws its inspiration directly from the concepts introduced by Activity Theory and more generally from the theories and methodologies introduced in previous sections.

The A&A perspective introduces in the context of intelligent agent systems a further dimension besides agent rationality, i.e. the dimension of the artifacts and tools to be conceived and designed to support agent rationality and activity. Apparently, this introduces a dualism between *Agens Faber* and *Agens Symbolicus*, analogous to the duality between Homo Faber or Homo Symbolicus [4] – who comes first? [19]. Such a dualism obviously has to be solved without a winner: then, why should we choose between an Agens Faber and an Agens Symbolicus while we aim

---

[2] Theoretically, by modifying the description of state to allow a single action selection at each decision point reduces the complexity of search from $b^n$ to $n$, where $n$ is the average depth to the goal node and $b$ the average branching factor

[3] Kirsh proposed here a Tetris experiment, showing how the possibility of rotating zoids in the game takes less time than rotating them mentally.

at intelligent agents? Accordingly, adopting an evolutionary perspective over agents, and carrying along the analogy with the development and evolution of human intelligence, we claim here that a theory of agent intelligence should not be limited to the modelling of inner, rational processes (as in BDI theory), and should instead include not only the basics of practical reasoning, but also a suitable theory of the artifacts and the means for their rational use, selection, construction and manipulation. This is in fact the idea behind the *Agens Faber* notion: agent intelligence should not be considered as separated by the agent's ability to perceive and affect the environment - and so, that agent intelligence is strictly related to the artifacts that enable, mediate and govern any agent (intelligent) activity.

Along this line, in the remainder of this section we first collect some considerations of ours about the conceptual relation between agents and artifacts, then we briefly describe a first model and taxonomy of artifacts introduced by the A&A approach.

### 4.1   On the Relation between Agents and Artifacts

**Goals of Agents and Use of Artifacts**  By considering the conceptual framework described in [6], agents can be generally conceived as *goal-governed* or *goal-oriented* systems. Goal-governed systems refer to the strong notion of agency, i.e. agents with some form of cognitive capabilities, which make it possible to explicitly represent their goals, driving the selection of agent actions. Goal-oriented systems refer to the weak notion of agency, i.e. agents whose behaviour is directly designed and programmed to achieve some goal, which is not explicitly represented. In both goal-governed and goal-oriented systems, goals are *internal*. *External goals* instead refer to goals which typically belong to the social context or environment where the agents are situated. External goals are sorts of regulatory states which condition agent behaviour: a goal-governed system follows external goals by adjusting internal ones.

This basic picture is then completed by systems which are not goal-oriented. This is the case of passive objects, which are characterised by the concept of *use*: they have not internal goals, but can be *used* by agents to achieve their goals. *Artifacts* are objects explicitly designed to provide a certain *function*[4], which guides their use. The concept of *destination* is related but not identical to the concept of function: it is an external goal which can be attached to an object or an artifact by users, in the act of using it. Then an artifact can be used according to a destination which is different from its function.

An interesting distinction has been proposed, concerning agents / artifacts relationships, between *use* and *use value* [6]: there, use value corresponds to the evaluation of artifact characteristics and function, in order to *select* it for a (future) use. The distinction corresponds to two different kinds of external goals attached to an artifact: *(i)* the use-value goal, according to which the artifact should allow user

---

[4] The term "function" here refers to the functionality embodied by an artifact, and should not be confused with the same term as used e.g. in mathematics or in programming languages

agents to achieve their objective, such an external goal drives the agent selection of the artifact; *(ii)* the use goal, which directly corresponds to the agent internal goal, which guides the actual usage of the artifact. From the agent point of view, when an artifact is selected and used it has then a use-value goal that somehow matches its internal goal.

By extending the above considerations, the classical tool-using / tool-making distinction from anthropology can be articulated along three main distinct aspects, which characterise the relationship between agents and artifacts:

- use
- selection
- construction and manipulation

While the first two aspects are clearly related to use and use value, respectively, the third is the rational consequence of a failure in the artifact selection process, or in the use of a selected artifact. Then, a new, different artifact should be constructed, or obtained by manipulation of an existing one.

**Agents Reasoning about Artifacts**  One of the key issues in the Agens Faber approach is how artifacts can be effectively exploited to improve agent ability to achieve individual as well as social goals. The main questions to be answered are then: How should agents reason to use artifacts in the best way, making their life simpler and their action more effective? How can agents reason to select artifacts to use? How can agents reason to construct or adapt artifact behaviour in order to fit their goals?

On the one hand, the simplest case concerns agents directly programmed to use specific artifacts, with usage protocols directly defined by the programmer either as part of the procedural knowledge / plans of the agent for goal-governed systems, or as part of agent behaviour in goal-oriented systems. In spite of its simplicity, this case can bring several advantages for MAS engineers, exploiting separation of concerns for programming simpler agents, by charging some burden upon specifically-designed artifacts. On the other hand, the intuition is that in the case of fully-open systems, the capability of the artifact to describe itself, its function, interface, structure and behaviour could be the key for building open MAS where intelligent agents dynamically look for and select artifacts to use, and then exploit them for their own goals.

At a first glance, it seems possible to frame the agent's ability to use artifacts in a hierarchy, according to five different cognitive levels at which the agent can use an artifact:

**unaware use** — at this level, both agents and agent designers exploit artifacts without being aware of it: the artifact is used implicitly, since it is not denoted explicitly. In other words, the representation of agent actions never refer explicitly to the execution of operation on some kind of artifacts.

**embedded / programmed use** — at this level, agents use some artifacts according to what has been explicitly programmed by the designer: so, the artifact selection is explicitly made by the designer, and the knowledge about its use is implicitly encoded by the designer in the agent. In the case of cognitive agents, for instance, agent designers can specify usage protocols directly as part of the agent plan. From the agent point of view, there is no need to understand explicitly artifact operating instructions or function: the only requirement is that the agent model adopted could be expressive enough to model in some way the execution of external actions and the perception of external events.

**cognitive use** — at this level, the agent designer directly embeds in the agent knowledge about what artifacts to use, but how to exploit the artifacts is dynamically discovered by the agent, reading the operating instructions. artifact selection is still a designer affair, while how to use it is delegated to the agent's rational capabilities. So, generally speaking the agent must be able to discover the artifact function, and the way to use it and to make it fit the agent goals. An obvious way to enable agent discovery is to make artifacts explicitly represent their function, interface, structure and behaviour.

**cognitive selection and use** — at this level, agents autonomously select artifacts to use, understand how to make them work, and then use them: as a result, both artifact selection and use are in the hands of the agents. It is worth noting that such a selection process could also concern sets of cooperative agents, for instance interested in using a coordination artifact for their social activities.

**construction and manipulation** — at this level, agents are lifted up to the role of designers of artifacts. Here, agents are supposed to understand how artifacts work, and how to adapt their behaviour (or to build new ones from scratch) in order to devise out a better course of actions toward the agent goals. Because of its complexity, this level more often concerns humans: however, not-so-complex agents can be adopted to change artifact behaviour according to some schema explicitly pre-defined by the agent designers.

## 4.2    A Model of Artifacts for MAS

In order to allow for its rational exploitation by intelligent agents, an artifact for MAS possibly exposes *(i)* a *usage interface*, *(ii)* *operating instructions*, and *(iii)* a *service description*. On the one hand, this view of artifacts provides us with a powerful key for the interpretation of the properties and features of existing non-agent MAS abstractions, which can be then catalogued and compared based on some common criteria. On the other hand, it is also meant to foster the conceptual grounding for a principled methodology for the engineering of MAS environment, where artifacts play the role of the core abstractions.

**Usage Interface** — One of the core differences between artifacts and agents, as computational entities populating a MAS, lays in the concept of *operation*, which is the means by which an artifact provides for a service or function. An agent executes an action over an artifact by invoking an artifact operation. Execution possibly terminates with an *operation completion*, typically representing the outcome of the invocation, which the agent comes to be aware of in terms of perception. The set of operations provided by an artifact defines what is called its *usage interface*, which (intentionally) resembles interfaces of services, components or objects, in the object-oriented sense of the term.

In MAS, this interaction schema is peculiar to artifacts, and makes them intrinsically different from agents. While an agent has no interface, acts and senses the environment, encapsulates its control, and brings about its goals proactively and autonomously, an artifact has instead a usage interface, is used by agents (and never the opposite), is driven by their control, and automatises a specific service in a predictable way without the blessing of autonomy. Hence, owning an interface strongly clearly differentiates agents and artifacts, and is therefore to be used by the MAS engineer as a basic discriminative property between them.

**Operating Instructions** — Coupled with a usage interface, an artifact could provide agents with *operating instructions*. Operating instructions are a description of the procedure an agent has to follow to meaningfully interact with an artifact over time. Most remarkably, one such description is history dependent, so that actions and perceptions occurring at a given time may influence the remainder of the interaction with the artifact. Therefore, operating instructions are basically seen as an exploitation protocol of actions / perceptions. This protocol is possibly furthermore annotated with information on the intended preconditions and effects on the agent mental state, which a rational agent should read and exploit to give a meaning to operating instructions. Artifacts being conceptually similar to devices used by humans, operation instructions play a role similar to a manual, which a human reads to know how to use the device on a step-by-step basis, and depending on the expected outcomes he/she needs to achieve. For instance, a digital camera provides buttons and panels (representing its usage interface), and therefore comes with a manual describing how to use them, e.g. which sequence of buttons are to be pushed to suitably configure the camera resolution.

**Function Description** — Finally, an artifact could be characterised by a *function description* (or service description). This is a description of the functionality provided by the artifact, which agents can use essentially for artifact selection. In fact, differently from operating instructions, which describe *how* to exploit an artifact, function description describes *what* to obtain from an artifact. Clearly, function description is an abstraction over the actual implementation of the artifact: it hides inessential details over the implementation of the service while highlighting key functional (input/output) aspects of it, to be used by agents for artifact selection. For instance, when modelling a sensor wrapper as an artifact, we may easily think of the operations for sensor activation and inspection as described via usage interface and operations instructions, while the information about the sensory function itself being conveyed through function description of the sensor wrapper.

Besides this model, some basic properties and features can be identified for artifacts, which possibly enhance agent ability to use them for their own purposes:

**Inspectability** — The state of an artifact, its content (whatever this means in a specific artifact), its usage interface, operating instructions and function description might be all or partially available to agents through *inspectability*. Whereas in closed MASs this information could be hard-coded in the agent - the artifact engineer develops the agents as well - in open MASs third-party agents should be able to dynamically join a society and get aware at run-time of the necessary information about the available artifacts. Also, artifacts are often in charge of critical MAS behaviour [17]: being able to inspect a part or the whole of an artifact features and state is likely to be a fundamental capability in order to understand and govern the dynamics and behaviour of a MAS.

**Controllability** — Controllability is an obvious extension of the inspectability property. The operational behaviour of an artifact should then not be merely inspectable, but also controllable so as to allow engineers (or even intelligent agents) to monitor its proper functioning: it should be possible to stop and restart an artifact working cycle, to trace its inner activity, and to observe and control a step-by-step execution. In principle, this would largely improve the ability of monitoring, analysing and debugging at execution time the operational behaviour of an artifact, and of the associated MAS social activities as well.

**Malleability** — Also related to inspectability, malleability (also called *forgeability*) is a key-feature in dynamic MAS scenarios, when the behaviour of artifacts could be required to be modified dynamically in order to adapt to the changing needs or mutable external conditions of a MAS. Malleability, as the ability to change the artifact behaviour at execution-time, is seemingly a crucial aspect in on-line engineering for MASs, and also a perspective key issue for self-organising MASs.

**Predictability** — Differently from agents, which as autonomous entities have the freedom of behaving erratically, e.g. neglecting messages, usage interface, operating instructions and function description can be used as a contract with an artifact by an agent. In particular, function description can provide precise details of the outcomes of exploiting the artifact, while operating instructions make the behaviour of an artifact predictable for an agent.

**Formalisability** — The predictability feature can be easily related with *formalisability*. Due to the precise characterisation that can be given to an artifact's behaviour, until reaching a full operational semantics model, for instance, as developed for coordination artifacts in [20], it might be feasible to automatically verify the properties and behaviour of the services provided by artifacts, for this is intrinsically easier than services provided by autonomous agents.

**Linkability** — artifacts can be used to encapsulate and model reusable services in a MAS. To scale up with complexity of an environment, it might be interesting to compose artifacts, e.g. to build a service incrementally on top of another, by making a new artifact realising its service by interacting with an existing artifact. To this end, artifacts should be able to invoke the operation of another artifact: the reply to that invocation will be transmitted by the receiver through the invocation of another operation in the sender.

**Distribution** — Differently from an agent, which is typically seen as a point-like abstraction conceptually located to a single node of the newtwork, artifacts can also be distributed. In particular, a single artifact can in principle be used to model a distributed service, accessible from more nodes of the net. Using link-ability, a distributed artifact can then be conceived and implemented as a composition of linked, possibly non-distributed artifacts, or vice versa, a number of linked artifacts, scattered through a number of different physical locations could be altogether seen as a single distributed artifact. Altogether, distribution and linkability promote the layering of artifact engineering, as sketched in Section 4.3.

As a final remark, it should be noted that all the artifact features presented above play a different role when seen from the different viewpoints of agents and of MAS engineers. For instance, operating instructions are mostly to be seen as a design tool for engineers, as well as a run-time support for rational agents. Instead, features like inspectability and malleability gain particular interest when the two viewpoints can be made one: when an intelligent agent is allowed to play and is capable of playing the role of the MAS engineer, it can in principle understand the state and dynamics of the MAS by observing the artifacts, then possibly working as an Agens Faber: that is, by re-working its tools (the artifacts) in order to suitably change the overall MAS behaviour.

### 4.3    A Basic Taxonomy of Artifacts for MAS

Many sorts of different artifacts can populate a MAS, providing agents with a number of different services, embodying a variety of diverse models, technologies and tools, and addressing a wide range of application issues. Correspondingly, a huge variety of approaches and solutions are in principle available for MAS engineers when working to shape the agent environment according to their application needs. So, the mere model of artifacts for MAS is no longer enough: a *taxonomy* of artifacts comes to be useful, which could help MAS engineers first defining the basic classes of artifacts, their differences and peculiarities, then classifying known artifacts, to understand and possibly compare them.

Among the many possible criteria for a taxonomy, we find it useful to focus on the mediation role of the artifact, and then discriminate artifacts based on the sort of (non-artifact) MAS entities they are meant to tie together. According to the pictorial representation in Fig. 1, our first proposal here divides artifacts into *individual artifacts*, *social artifacts*, and *resource artifacts*.

Individual artifacts are artifacts exploited by one agent only, in other terms, an individual artifact mediates between an individual agent and the environment. Individual artifacts can serve several purposes, including externally enhancing agent capabilities, such as e.g. adding a private external memory, enacting a filtering policy of the agent actions toward other artifacts (as in the case of agent coordination contexts [18]), providing individual agents with useful information on the organisation, and so on. In general, individual artifacts are not directly affected by the activity of other agents, but can, through linkability, interact with other artifacts in the MAS.

Social artifacts are instead artifacts exploited by more than one agent. In other terms, a social artifact mediates between two or more agents in a MAS. In general, social artifacts typically provide a service which is in the first place meant to achieve a social goal of the MAS, rather than an individual agent goal. For instance, social artifacts might provide a coordination service [25], governing the activities of two or more agents, as for example in multi-party protocols, but can also realise global knowledge repositories, shared ontologies, or organisation abstractions containing information on roles and permissions.

Finally, resource artifacts are artifacts that conceptually wrap external resources, in other terms, a resource artifact mediates between a MAS and an external resource. External resources can be either legacy components and tools, applications written with non-agent technologies because of engineering convenience, such as Web Services, or physical resources which the agents of a MAS might need to act upon and sense. In principle, resource artifacts can be conceived as a means to raise external MAS resources up to the agent cognitive level. In fact, they provide external resources with an usage interface, some operating instructions, and a service description, and realise their task by dynamically mapping high-level agent interactions upon lower-level interactions with the resources, using e.g. specific transports such as object-oriented local or remote method calls, HTTP requests, and the like.

Altogether, individual, social and resource artifacts can be used as the basis for building the glue keeping agents together in a MAS, and for structuring the environment where agents live and interact. In fact, our taxonomy, as apparent from Fig. 1, defines a structured, *layered* view over the MAS environment, and implicitly suggests a model for organising agent interaction within a MAS. As such, the artifact taxonomy could lead to a well-principled foundation for a general agent-oriented methodology for the engineering of the agent environment as a first-class entity.

# 5 Conclusion

By drawing an analogy between intelligence in MAS and the development and evolution of human intelligence, in this chapter we elaborated on MAS environment, agent tools, and their relationship with agent intelligence. Our metaphor of the Agens Faber comes to say that a theory of agent intelligence should not be limited to modelling the inner rational process of an agent, but should instead include not only the basics of practical reasoning, but also a theory of the agent *artifacts*, as defined in the A&A meta-model for MAS, providing agents with the means for the rational use, selection, construction, and manipulation of artifacts.

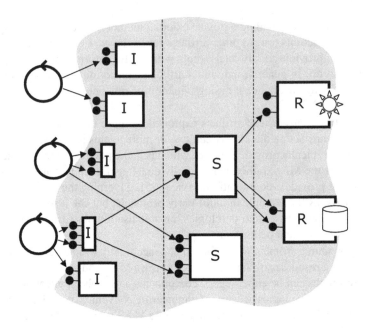

**Fig. 1.** Individual, social, and resource artifacts: a layered view of artifacts for MAS, from [19].

# References

1. Agre, P.E., Horswill, I.: Lifeworld analysis. Journal of AI Research 6(1), 111–145 (1997)
2. St. Amant, R., Wood, A.B.: Tool use for autonomous agents. In: Proceedings of the National Conference on Artificial Intelligence, AAAI-05 (2005)
3. Beck, B.B.: Animal tool behavior. the use and manufacture of tools by animals. The Quarterly Review of Biology 56(2), 231 (1981)
4. Berggren, K.: Homo Faber or Homo Symbolicus? The fascination with copper in the sixth millennium. Transoxiana: Journal Libre de Estudios Orientales 8 (2004)
5. Clark, A., Chalmers, D.: The extended mind. Analysis 58(1), 7–19 (1998)
6. Conte, R., Castelfranchi, C. (eds.): Cognitive and Social Action. University College London (1995)
7. Gibson, K.R., Ingold, T.: Tools, Language and Cognition in Human Evolution. Cambridge University Press, Cambridge (1995)
8. Hewes, G.W.: A history of speculation on the relation between tools and languages. In: Tools, Language and Cognition in Human Evolution, pp. 20–31. Cambridge University Press, Cambridge (1993)
9. Hutchins, E.: Cognition in the Wild. MIT Press, Cambridge (1995)
10. Kirsh, D.: The intelligent use of space. Artificial Intelligence 73(1-2), 31–68 (1995)
11. Leont'ev, A.N.: The making of mind: A personal account of Soviet psychology. Harvard University Press, Cambridge (1979)
12. Leont'ev, A.N.: Problems of the development of mind. Progress Publishers, Moscow (1981)

13. Maes, P.: Agents that reduce work and information overload. Commun. ACM 37(7), 30–40 (1994)
14. Martelet, G.: Evolution et Creation, tome 1. Paris (1998)
15. Newell, A., Simon, H.A.: Human Problem Solving. Prentice-Hall, Englewood Cliffs (1972)
16. Norman, D.A.: Cognitive artifacts. In: Carroll, J. (ed.) Designing interaction: psychology at the human-computer interface. Cambridge Series On Human-Computer Interaction archive, pp. 17–38. Cambridge University Press, Cambridge (1991)
17. Omicini, A., Ossowski, S., Ricci, A.: Coordination infrastructures in the engineering of multiagent systems. In: Bergenti, F., Gleizes, M.-P., Zambonelli, F. (eds.) Methodologies and Software Engineering for Agent Systems: The Agent-Oriented Software Engineering Handbook, pp. 273–296. Kluwer Academic Publishers, Dordrecht (2004)
18. Omicini, A., Ricci, A., Viroli, M.: Formal specification and enactment of security policies through Agent Coordination Contexts. Electronic Notes in Theoretical Computer Science 85(3) (2003), 1st International Workshop "Security Issues in Coordination Models, Languages and Systems" (SecCo 2003), Eindhoven, The Netherlands, 28–29 June 2003, Proceedings
19. Omicini, A., Ricci, A., Viroli, M.: Agens Faber. Toward a theory of artefacts for MAS 150(3), 21–36 (2006), 1st International Workshop "Coordination and Organization" (CoOrg 2005), COORDINATION 2005, Namur, Belgium, 22 April 2005, Proceedings
20. Omicini, A., Ricci, A., Viroli, M., Castelfranchi, C., Tummolini, L.: Coordination artifacts: Environment-based coordination for intelligent agents. In: Jennings, N.R., Sierra, C., Sonenberg, L., Tambe, M. (eds.) 3rd International Joint Conference on Autonomous Agents and Multiagent Systems (AAMAS 2004), New York, NY, USA, 19–23 July 2004. 1, pp. 286–293. ACM Press, New York (2004)
21. Povinelli, D.J.: Folk Physics for Apes: The Chimpanzees Theory of How the World Works. Oxford University Press, Oxford (2000)
22. Ricci, A., Viroli, M., Omicini, A.: "Give agents their artifacts": The A&A approach for engineering working environments in MAS. In: Durfee, E., Yokoo, M., Huhns, M., Shehory, O. (eds.) 6th International Joint Conference "Autonomous Agents & Multi-Agent Systems" (AAMAS 2007), Honolulu, Hawaii, USA, 14–18 May 2007, pp. 14–18. IFAAMAS (2007)
23. Simon, H.A.: The Sciences of the Artificial. MIT Press, Cambridge (1981)
24. Sterenly, K.: The Externalist Challenge: New Studies on Cognition and Intentionality. Walter de Gruyter, Berlin (2003)
25. Viroli, M., Omicini, A.: Coordination as a service. Fundamenta Informaticae (Special Issue: Best papers of FOCLASA 2002) 73(4), 507–534 (2002)
26. Vygotsky, L.S.: Mind in Society: The Development of Higher Psychological Processes. Harvard University Press, Cambridge (1980)
27. Wertsch, J.V.: Vygotsky and the social formation of mind. Harvard University Press, Cambridge (1979)
28. Weyns, D., Van Dyke Parunak, H. (eds.): Journal of Autonomous Agents and Multi-Agent Systems (Special Issue: Environment for Multi-Agent Systems) 14(1) (2007)
29. Wundt, W.: Elements of folk psychology: Outlines of a psychological history of the development of mankind. Allen and Unwin, Crows Nest (1916)

# An Overview of AI Research in Italy

Andrea Roli[1] and Michela Milano[2]

[1] DEIS, Campus of Cesena, *Alma Mater Studiorum* Università di Bologna, via Venezia 52,
I-47023 Cesena, Italy, andrea.roli@unibo.it
[2] DEIS, *Alma Mater Studiorum* Università di Bologna, viale Risorgimento 2,
I-40136 Bologna, Italy, michela.milano@unibo.it

**Abstract.** This chapter aims to provide an overview of the main Italian research areas and activities. We first analyze the collaboration structure of Italian research, which involves more than eight hundred scholars and researchers from both universities and industry. From a network perspective it appears to be scale-free. Next, we briefly illustrate the main subjects of investigation and applications. AI research in Italy goes back to the 1970s with an increase in the last twenty years and spans the main research AI areas, from automated reasoning and ontologies to machine learning, robotics and evolutionary computation.

## 1  Introduction

The 50th anniversary of the 1956 Dartmouth Conference was an occasion for AI research communities to look back on those fifty years of research, make a synthesis of the state of the art and consider future work. So it was in Italy, where the Italian Association for Artificial Intelligence (AI*IA) published a special issue on AI research state of the art, with particular emphasis on Italian research [123].

The special issue is the starting point of this chapter, aimed at providing an overview of the main Italian research areas and activities. AI research in Italy traces back to the 1970s with an increase in the last twenty years and spans the main research AI areas, from automated reasoning and ontologies to machine learning and robotics. In fact, the areas of interest are quite numerous and can be classified as follows:

- Knowledge representation and reasoning
- Constraint Satisfaction and Optimization
- Planning and Scheduling
- Automated diagnosis
- AI and entertainment
- Machine learning and data mining
- Kernel machines, neural networks and graphical models
- Multiagent systems
- Robotics
- Genetic and evolutionary computation
- Complex systems

M. Bramer (Ed.): Artificial Intelligence, LNAI 5640, pp. 174–192, 2009.
© IFIP International Federation for Information Processing 2009

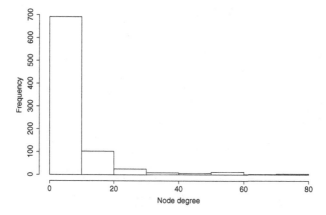

**Fig. 1.** Node degree frequency of Italian collaboration network in AI.

A rigid classification obviously introduces some inaccuracies and margins for arbitraries, especially because the borders between the areas are quite blurred. However, this list of AI research topics still is a representative description of AI domains of Italian research.

A general characteristic of Italian research (in AI) is that it is uniformly distributed throughout the country and the collaborations between universities and other institutions are quite tight. Moreover, the interaction with foreign research groups, both in Europe and outside it, is very lively. A selection of Italian research in AI is provided by the proceedings of the biannual conference organized by AI*IA, published by Springer in Lecture Notes in Artificial Intelligence series.

A picture of the publication and collaboration networks among AI Italian researchers and between Italian and foreign groups can be drawn by analyzing the properties of a representative sample of the collaboration graph built by analyzing the publications collected in the DBLP database [86].[3] We studied the graph resulting from a representative sample of papers published by Italian research groups in AI and observed that the different Italian authors are more than eight hundred in number and the foreign authors involved as co-authors are about four hundred. We also analyzed the collaboration network, restricted to Italian authors only, defined as the undirected graph in which nodes correspond to authors and two nodes are connected if the two corresponding authors had at least one joint publication. The frequency of node degree is plotted in Figure 1. Very interestingly, but not surprisingly, the collaboration network seems to be scale-free, that is, the node degree is described by a power-law distribution: roughly, few authors have a huge number of links, while many have just a few [22]. Indeed, in Figure 1 we can observe that the points, plotted in log-log scale, can be fitted by a line with slope

---

[3] We parsed and processed the XML file containing the publications collected in DBLP at the time of writing (October, 2007).

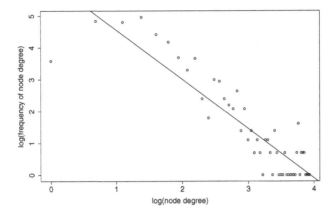

**Fig. 2.** Log transformation of node degree frequency of Italian collaboration network in AI, fitted with a line with slope equal to -1.553.

equal to -1.553. Hence, the frequency of node degree $k$ is ruled by the following law: $\mathrm{freq}(k) \sim k^{-1.553}$. This result is in accordance with previous results [23] and suggests that scientific collaboration in Italy is governed by the same dynamics as international collaboration networks.

In the remainder of this chapter we briefly describe the main contributions of Italian research groups to each of the AI areas listed above. For each domain a representative selection of works is mentioned; however, the aim of this overview is not to provide a complete list of topics and publications, but rather to draw a high level picture of Italian AI research. Moreover, since we are obviously not immune from human fragilities, notwithstanding our efforts in trying to cite all the main Italian scholars and researchers, some might still be missing and we apologize to them and to our readers in case of such an accident.

## 2   Knowledge Representation and Reasoning

A large fraction of Italian research in AI is devoted to logic-based knowledge representation and automated reasoning. Since the 1980s, there is a strong interest in logics for AI, especially on logic formalisms [141], modal logics [117], meta-reasoning [81,57], and spatial logic [3]. The use of logics has several advantages: it is declarative and high level, it has a formal semantics, has an embedded reasoning apparatus and it is intuitive.

The interest in the *satisfiability problem* (SAT) and its extensions, such as *Quantified boolean formulas* (QBF), is quite strong, as proved by the active research in SAT and QBF solvers [116,11,27].

Italian researchers also strongly contributed in the development of research into ontologies, since the late 1980s. The contributions are in two directions: one with focus on knowledge representation languages with predictable computational behaviour and sound and complete inference algorithms [113,53]; the second aiming at addressing ontology in a systematic and principled way [173,121]. The studies on ontologies have been largely applied to the semantic web [175,101]. The activeness of research in the field is also witnessed by the role played by the Italian community in the *Applied Ontology* journal [8], published by IOS Press, that includes Italian researchers on the editorial board, as well as in co-editing.

An important contribution of ontologies to the strictly related field of natural language processing (NLP) can be found in [106], in which an upper level ontology named DOLCE (Descriptive Ontology for Linguistic and Cognitive Engineering) is proposed. Contributions in NLP from Italian researchers can also be found in [9,4,24,25]. For a detailed survey of Italian contributions to research in ontologies, we forward the interested reader to [54].

As far as automated reasoning is concerned, the Italian community has contributed to theoretical and practical achievements. In [110] the definition of a generic theorem prover is presented. The study of formalisms for automated reasoning as a search problem is proposed in [41] while search strategies for SAT solvers are outlined in [40]. Tractable reasoning problem classes are surveyed in [48] along with methods for automated model building. Tableau methods for non-monotonic reasoning have been investigated and surveyed in [150].

The current scenario shows very active research in extensions of logic programming, such as answer set programming [131,89,82,51] abductive logic programming [126], disjunctive logic programming[52], many-valued logic [7], temporal action logic [115]and model checking [29,97].

Automated reasoning finds widespread application in the field of multiagent systems programming and verification [6,16,164,15,83], bioinformatics [152], security protocols [58], web systems and applications [31,5]. In particular, web services and their composition is a very active area of application [138,122].

## 3   Constraint Satisfaction and Optimization

Constraint satisfaction and optimization was born as a rib of the AI area of knowledge representation and reasoning but it has been influenced by other areas such as programming languages, operations research, graph theory and software engineering. Nowadays all the concepts studied in this field are integrated into Constraint Programming (CP) languages. CP is now an established powerful paradigm for solving hard combinatorial search/optimization problems [139]. Italian research on constraint satisfaction dates back to the early seventies with the studies by Ugo Montanari on picture processing [145]. Nowadays, Italian research is mostly divided into a stream oriented towards theoretical aspects and solvers and a stream aimed at using constraint based techniques for solving real life applications.

As far as theoretical aspects are concerned, properties of Constraint Logic Programming languages are studied using partial evaluation [99] and abstract interpretation [13]. Concurrent constraint programs have been investigated in Italy, by defining extensions of the standard programming paradigm where notions of time can be modelled [39].

With regard to solvers, Italian research has been devoted to studying constraint solvers on sets/multi-sets [90], *soft constraint* solvers [36,35] including several constraint classes, such as fuzzy, possibilistic, probabilistic and weighted, solvers integrating operations research techniques in CP [144], and solvers integrating metaheuristics and stochastic local search in a constraint based framework [38,108,107].

On the application side, many fields have been investigated by Italian researchers: planning and scheduling (see Section 4), temporal reasoning [158,163], security and quality of service [37], bioinformatics [153], embedded systems-on-chip design [30], timetabling [142], system verification [87].

## 4    Planning and Scheduling

The design of systems able to make plans autonomously to accomplish a target task and to allocate actions and resources efficiently while satisfying given constraints is one of the most notable goals of AI. Research in planning and scheduling is quite vigorous since the 1970s and the main Italian contribution to the field can be found in works addressing the issue of designing efficient planning and scheduling techniques. This goal is achieved especially by exploiting alternative formulations, such as model checking and SAT, in the case of planning; whereas, in the case of scheduling advanced metaheuristic techniques are used. Representative examples of these researches can be found in [14,62,74,28] in which planning problems are mainly tackled by reformulations. CP and metaheuristics can also be used effectively to solve both planning and scheduling problems, as discussed respectively in [66] and [109].

## 5    Automated Diagnosis

The main contribution of Italian research in automated diagnosis can be found in advances in model-based diagnosis, that face the problem of providing fault diagnosis by using one or more models of the system to be analyzed. Among the main contributions brought from Italian research we mention works on different kinds of modeling, such as causal models [75] and process algebras [77]. In general, the problem is computationally hard and in recent years a number of approaches have been investigated for making the diagnosis process more efficient. Among these approaches, the one that makes use of multiple models has been deeply studied by Italian researchers [72,178,160,42]. A significant contribution is the one in [71], in which a simpler version of the diagnosis problem is automatically derived by exploiting the amount of current observations on the problem.

Off-line vs. on-line diagnosis has been deeply studied, in a collaboration between the University of Torino and *Centro Ricerche Fiat* (Fiat Research Center); in particular, the problems arising in embedding diagnostic software in vehicles equipped with on-board electronic control units have been tackled in [60,76].

A detailed overview of Italian research in the field, along with a rich bibliography, can be found in [78].

## 6   AI and Entertainment

One of the most active research areas in AI is the one comprising game playing, intelligent and affective user interfaces, collaborative and virtual environments. Italian contributions to this field are in games, entertaining communication and educational entertainment.

A representative work in game playing is *WebCrow* [91], a crossword puzzle solver that is composed of a web-query and information processing module for retrieving candidate answers to crossword clues, and a probabilistic-CSP module that tries to fill the crossword grid in the best way. One of the strengths of *WebCrow* is that it is designed to be independent of the language used for the puzzles.

Entertaining communication characterizes intelligent multimedia interfaces and tries to reproduce the peculiarities of human communication, such as humour and storytelling. Italian research contributed to this topic with the HAHAcronym project [171], whose goal is to develop a system able to produce humorous versions of existing acronyms or, dually, to create amusing acronyms from a given vocabulary. which formalize the origins of humor as unexpected meanings or behaviors.

In the field of educational entertainment, Italy made an important contribution with AI systems for museum guides with the PEACH system [172] that comprises seamlessly integrated mobile and stationary components and in which the user has a personalized profile and the guide dynamically adapts to the user preferences.

## 7   Machine Learning and Data Mining

Italian research in Machine learning (ML), in its traditional meaning, was initially pioneered by researchers from the Universities of Torino and Bari, joined also by personnel from the *Ugo Bordoni* Foundation, IBM Italia and the Telecom research lab (TiLab). This core group then grew and many other Italian universities contributed to research in ML. Italian research in ML now forms a dense collaboration network, with many links to foreign countries and it is present in the most important *fora* of the subject. A description of this historical aspect, along with a detailed bibliography, can be found in [92].

The main contributions can be found in concept learning for classification tasks, both with supervised approaches [94,136,32,169,140] and unsupervised approaches (namely, clustering), such as [59] and multiple classifier systems [103]. Important contributions can also be found in computational learning theory [65], complexity of learning [44,114], changes of representation and incremental methods [165,95] and probabilistic knowledge and logic programming [161].

Applications of ML have also received the attention of the Italian ML community. ML has been applied to the semantic web [133], for the classification of web documents [64], for intelligent processing of printed documents [93] and for intelligent searching in digital libraries [132].

On the data mining side, we mention works on the extraction of association rules from databases [134], on handling geographical information [135], automatically building models of sequences [43,104] and data visualization [127].

# 8   Kernel Machines, Neural Networks and Graphical Models

Italian research in statistical and probabilistic learning is distributed in many theoretical and applicative fields, from the issue of bridging statistical learning and symbolic approaches, to bioinformatics and image recognition. An important contribution to statistical and probabilistic learning applied in structured domains can be found in [102] in which a framework for unifying artificial neural networks and belief nets in the context of structured domains is proposed. Research is also quite active in the combination of statistical and symbolic learning, such as kernel methods with Inductive Logic Programming [128] or with Prolog trees [154].

Important contributions come also from studies in recursive neural networks. For example, in [34] the authors define a methodology that makes it possible to feed recursive networks with cyclic graphs, thus extending their applicability and establishing their computational power and in [119] sufficient conditions for guaranteeing the absence of local minima of the error function in the case of acyclic graphs are presented. The studies reported in [170,143] address the issue of applying recursive networks on structured data.

As for applications of kernel methods and artificial neural networks, we mention works in bioinformatics such as the ones in [184,183,33,159] in which neural networks and kernel methods are successfully applied in several biological and chemical contexts, and contributions in document analysis such as [118] and references therein. Finally, we mention a work describing a neuro-symbolic language for monotonic and non-monotonic parallel logical inference by means of artificial neural networks [45].

# 9   Multiagent Systems

Research in multiagent systems (MAS) spans many different AI areas, from Distributed AI to robotics and programming languages and also important related disciplines such as sociology, ethology, biology and economics. In the last ten years, Italian (and European) researchers strongly contributed to research in MAS by producing important advances on many fronts. Notably, the AI*IA working group on MAS is very active and counts hundreds of participants from Italian and European universities and industries [80].

The initial steps of research in MAS were aimed at designing intelligent agent architectures and defining, first, proper communication languages, then infrastructures and development tools. A remarkable contribution from Italy is JADE [26], nowadays the most used agent-oriented platform. Coordination plays a crucial role in MAS and Italian researchers made a contribution to this subject [46,73] and produced coordination technologies, from field-based [137] to tuple-based [147,151] control and coordination infrastructures [84]. Moreover, Italian scholars contribute to MAS research from the perspective of cognitive science, e.g., [157,96] and artificial institutions, see, for instance, [100].

Applications of MAS are widely spread among many real-life sectors and Italian research groups produced relevant or prominent contributions to a large number of applications. For instance, in information management [47,105], health care [10], decision support systems [155] and bioinformatics [12].

MAS is also strictly related, and often interwoven, with other research areas, such as simulation of social and biological systems. Along these research lines, representative Italian contributions can be found in [17] –in which crowd behavior is modeled by means of cellular automata– and, in [56], that presents a MAS framework for systems biology. Relevant contributions in the simulation and analysis of social systems can be found in [61] and works by the same author.

## 10 Robotics

In this section we outline the major contribution of Italian researchers in the field of AI robotics. First of all, Italian research groups have regularly taken part in international robotics competitions since the 1990s. Good results have been achieved in the RoboCup competitions: in the Middle-size league with ART (Azzurra Robot Team) [146], developed by teams from several universities and the Consorzio Padova Ricerche, and in the Humanoid Robot league with the humanoid robot Isaac [125] designed by Politecnico of Torino. As is often the case, the quest for smart robots able to carry out complex tasks also brings important advances in theory and methodology. Indeed, the development of ART led to the definition of effective communication and coordination protocols [124].

Another Italian project to be mentioned is RoboCare [162], that is currently running and is aimed at designing intelligent robots for human assistance.

One of the problems related to robotics is artificial vision. In Italy, there are several groups studying this subject, see for example [156,67,69,85]. Artificial vision has been successfully applied to autonomous vehicles [179] and exploited in the TerraMax competition.

Relevant results in the field of AI robotics and swarm intelligence have been achieved by the group of Stefano Nolfi at CNR (Italian Research Center), especially in the context of European projects such as Swarm-bots [174]. Among the main contributions, we mention the work on evolutionary robotics and collective behaviour [148,180].

Finally, more in the line of foundational research, there are some notable works in action theories, such as the ones on situation calculus formalism [98], dynamic logics [112] and fluent and event calculi [70]. It is important also to mention works in which a way to bridge the gap between action theory and applications on real robots is proposed [63,68].

# 11  Genetic and Evolutionary Computation

Evolutionary computation (EC) is inspired by Darwin's theory of natural selection and evolution and it is one of the most prominent and successful examples of AI techniques inspired by biological phenomena and metaphors. Its biological roots connects it tightly to other fields such as Ant algorithms, Particle swarm and Artificial immune systems. Moreover, EC provides a set of methods and tools that can be applied effectively in many fields of AI, such as robotics, fuzzy logics and neural networks. Italian research in EC is distributed among all the facets of the field and involves universities and institutions all over the country. Several of the best-known researchers in EC and related fields are Italian. For instance, Riccardo Poli is a world leader in genetic programming [130], Marco Dorigo is the inventor of *ant colony optimization* [88] and Marco Tomassini is a very well known researcher in EC and complex systems [176].

A selection of Italian contributions to EC can be classified in the following areas: classification [49], image processing [50,149], cellular evolutionary algorithms and cellular automata [111,182], test pattern generation [79], genetic programming [177] and hybrid architectures [176].

# 12  Complex Systems

Complex systems sciences (CSS) include disciplines like physics, dynamical systems, computer science, biology and sociology and it overlaps with many areas of AI [168]. Besides core principles such as information processing, computation and learning, common themes between CSS and AI are neural networks, evolutionary computation, multiagent systems and robotics. Italian research in CSS is spread among these domains and involves both academia and industry. The most active topics are artificial life, systemic, simulation of multiagent systems, cognitive processes, evolutionary computation and neural networks. There are also important international events promoted by Italian researchers, such as the Agent Based Modeling and Simulation symposium (ABModSim) [1] and the International Conference on Cellular Automata and Industries (ACRI) [2]. A carefully compiled list of websites devoted to CSS research in Italy, along with further bibliographic references, can be found in [21].

One of the main contributions of Italian research in CSS & AI is surely to be found in simulation of multiagent, social and biological systems. In particular, topics related to cellular automata [20] have characterized Italian research since the 1970s and have been applied in many contexts, ranging from ecology [120], to rubber compounds [18], cellular biology [166], traffic [55] and crowd behaviour [19]. Moreover, we find important contributions also in genetic networks [167] and social-economic systems [129].

**Acknowledgement**  We thank Luigia Carlucci Aiello, Federico Chesani, Marco Gavanelli, Paola Mello and Fabrizio Riguzzi for helpful suggestions and comments.

# References

1. http://www.lintar.disco.unimib.it/ABModSim/ (Viewed October 2007)
2. http://acri2006.univ-perp.fr/ (Viewed October 2007)
3. Aiello, M., Pratt-Hartmann, I.E., van Benthem, J.F. (eds.): Handbook of Spatial Logics. Kluwer Academic Publishers, Dordrecht (2007)
4. Ajani, G., Lesmo, L., Boella, G., Mazzei, A., Rossi, P.: Terminological and ontological analysis of european directives: multilinguism in law. In: ICAIL, pp. 43–48 (2007)
5. Alberti, M., Chesani, F., Gavanelli, M., Lamma, E., Mello, P., Montali, M., Torroni, P.: Web service contracting: Specification and reasoning with SCIFF. In: Franconi, E., Kifer, M., May, W. (eds.) ESWC 2007. LNCS, vol. 4519, pp. 68–83. Springer, Heidelberg (2007)
6. Alberti, M., Chesani, F., Gavanelli, M., Lamma, E., Mello, P., Torroni, P.: Verifiable agent interaction in abductive logic programming: the SCIFF framework. ACM Transactions on Computational Logics (To appear)
7. Anantharaman, S., Bonacina, M.P.: An application of automated equational reasoning to many-valued logic. In: Okada, M., Kaplan, S. (eds.) CTRS 1990. LNCS, vol. 516, pp. 156–161. Springer, Heidelberg (1991)
8. http://www.applied-ontology.org/ (Viewed October 2007)
9. Ardissono, L., Lesmo, L., Pogliano, P., Terenziani, P.: Interpretation of definite noun phrases. In: IJCAI, pp. 997–1002 (1991)
10. Ardissono, L., Leva, A.D., Petrone, G., Segnan, M., Sonnessa, M.: Adaptive medical workflow management for a context-dependent home healthcare assistance service. Electr. Notes Theor. Comput. Sci. 146(1), 59–68 (2006)
11. Armando, A., Castellini, C., Giunchiglia, E., Giunchiglia, F., Tacchella, A.: SAT-based decision procedures for automated reasoning: A unifying perspective. In: Mechanizing Mathematical Reasoning, pp. 46–58 (2005)
12. Armano, G., Mancosu, G., Orro, A.: Using multiple experts for predicting protein secondary structure. In: Artificial Intelligence and Applications, pp. 451–456 (2005)
13. Bagnara, R., Giacobazzi, R., Levi, G.: Static analysis of CLP programs over numeric domains. In: Billaud, M., Castéran, P., Corsini, M., Musumbu, K., Rauzy, A. (eds.) Actes "Workshop on Static Analysis '92", Bigre, Bordeaux, pp. 43–50, Extended abstract (1992)
14. Baioletti, M., Marcugini, S., Milani, A.: Dpplan: An algorithm for fast solutions extraction from a planning graph. In: AIPS, pp. 13–21 (2000)
15. Baldoni, M., Baroglio, C., Martelli, A., Patti, V.: A priori conformance verification for guaranteeing interoperability in open environments. In: Dan, A., Lamersdorf, W. (eds.) ICSOC 2006. LNCS, vol. 4294, pp. 339–351. Springer, Heidelberg (2006)
16. Baldoni, M., Martelli, A., Patti, V., Giordano, L.: Programming rational agents in a modal action logic. Ann. Math. Artif. Intell. 41(2-4), 207–257 (2004)
17. Bandini, S., Federici, M.L., Vizzari, G.: A methodology for crowd modelling with situated cellular agents. In: WOA, pp. 91–98 (2005)
18. Bandini, S., Magagnini, M.: Parallel processing simulation of dynamic properties of filled rubber compounds based on cellular automata. Parallel Computing 27(5), 643–661 (2001)
19. Bandini, S., Manzoni, S., Vizzari, G.: Situated cellular agents: A model to simulate crowding dynamics. IEICE Transactions on Information and Systems E87-D(3), 669–676 (2004)

20. Bandini, S., Mauri, G., Serra, R.: Cellular automata: From a theoretical parallel computational model to its application to complex systems. Parallel Computing 27(5), 539–553 (2001)
21. Bandini, S., Serra, R.: Complex systems. Intelligenza Artificiale 3(1-2), 102–108 (2006)
22. Barabasi, A.L.: Linked: The New Science of Networks. Perseus Books Group, New York (2002)
23. Barabási, A.L., Jeong, H., Ravasz, R., Néda, Z., Vicsek, T., Schubert, A.: On the topology of the scientific collaboration network. Physica A 311, 590–614 (2002)
24. Basili, R., Moschitti, A., Pazienza, M.T.: Extensive evaluation of efficient nlp-driven text classification. Applied Artificial Intelligence 20(6), 457–491 (2006)
25. Basili, R., Moschitti, A., Pazienza, M.T., Zanzotto, F.M.: Personalizing web publishing via information extraction. IEEE Intelligent Systems 18(1), 62–70 (2003)
26. Bellifemine, F., Poggi, A., Rimassa, G.: Developing multi-agent systems with a fipa-compliant agent framework. Software Practice & Experience 31(2), 103–128 (2001)
27. Benedetti, M.: Abstract branching for quantified formulas. In: Proceedings, The Twenty-First National Conference on Artificial Intelligence and the Eighteenth Innovative Applications of Artificial Intelligence Conference, AAAI 2006, July 16-20 (2006)
28. Benedetti, M., Aiello, L.C.: Sat-based cooperative planning: A proposal. In: Hutter, D., Stephan, W. (eds.) Mechanizing Mathematical Reasoning. LNCS (LNAI), vol. 2605, pp. 494–513. Springer, Heidelberg (2005)
29. Benerecetti, M., Giunchiglia, F., Serafini, L.: Model checking multiagent systems. Journal of Logic and Computation 8(3), 401–423 (1998)
30. Benini, L., Bertozzi, D., Guerri, A., Milano, M.: Allocation and scheduling for MPSoCs via decomposition and no-good generation. In: IJCAI, pp. 1517–1518 (2005)
31. Berardi, D., Giacomo, G.D., Lenzerini, M., Mecella, M., Calvanese, D.: Synthesis of underspecified composite e-services based on automated reasoning. In: Aiello, M., Aoyama, M., Curbera, F., Papazoglou, M.P. (eds.) ICSOC, pp. 105–114. ACM Press, New York (2004)
32. Bergadano, F., Giordana, A., Saitta, L.: Machine Learning: A General Framework and is Applications. Ellis Horwood, New York (1991)
33. Bernazzani, L., Duce, C., Micheli, A., Mollica, V., Sperduti, A., Starita, A., Tiné, M.R.: Predicting physical-chemical properties of compounds from molecular structures by recursive neural networks. Journal of Chemical Information and Modeling 46(5), 2030–2042 (2006)
34. Bianchini, M., Gori, M., Sarti, L., Scarselli, F.: Recursive processing of cyclic graphs. IEEE Transactions on Neural Networks 17(1), 10–18 (2006)
35. Bistarelli, S., Foley, S.N., O'Sullivan, B.: Detecting and eliminating the cascade vulnerability problem from multilevel security networks using soft constraints. In: AAAI, pp. 808–813 (2004)
36. Bistarelli, S., Rossi, F.: Semiring-based contstraint logic programming: syntax and semantics. ACM Transactions of Programming Languages and Systems 23(1), 1–29 (2001)
37. Bistarelli, S., Foley, S.N., O'Sullivan, B.: Detecting and eliminating the cascade vulnerability problem from multi-level security networks using soft constraints. In: Proc. Innovative Applications of Artificial Intelligence Conference, IAAI-04 (2004)
38. Blum, C., Roli, A.: Metaheuristics in combinatorial optimization: Overview and conceptual comparison. ACM Computing Surveys 35(3), 268–308 (2003)
39. de Boer, F., Gabbrielli, M., Meo, M.: Proving correctness of timed concurrent constraint programs. ACM Transactions on Computational Logic 5(4) (2004)

40. Bonacina, M.P.: A taxonomy of theorem-proving strategies. In: Artificial Intelligence Today, pp. 43–84. Springer, Heidelberg (1999)
41. Bonacina, M.P., Hsiang, J.: On the modelling of search in theorem proving - towards a theory of strategy analysis. Inf. Comput. 147(2), 171–208 (1998)
42. Bonarini, A., Sassaroli, P.: Opportunistic multimodel diagnosis with imperfect models. Inf. Sci. 103(1-4), 161–185 (1997)
43. Botta, M., Galassi, U., Giordana, A.: Learning complex and sparse events in long sequences. In: ECAI, pp. 425–429 (2004)
44. Botta, M., Giordana, A., Saitta, L., Sebag, M.: Relational learning as search in a critical region. Journal of Machine Learning Research 4, 431–463 (2003)
45. Burattini, E., de Francesco, A., Gregorio, M.D.: Nsl: A neuro-symbolic language for a neuro-symbolic processor (nsp). Int. J. Neural Syst. 13(2), 93–101 (2003)
46. Busi, N., Ciancarini, P., Gorrieri, R., Zavattaro, G.: Coordination models: A guided tour. In: Omicini, A., Zambonelli, F., Klusch, M., Tolksdorf, R. (eds.) Coordination of Internet Agents: Models, Technologies, and Applications, pp. 6–24. Springer, Heidelberg (2001)
47. Cabri, G., Guerra, F., Vincini, M., Bergamaschi, S., Leonardi, L., Zambonelli, F.: MOMIS: Exploiting agents to support information integration. Int. J. Cooperative Inf. Syst. 11(3), 293–314 (2002)
48. Caferra, R., Leitsch, A., Peltier, N.: Automate Model Building. Kluwer Academic Publishers, Dordrecht (2004)
49. Cagnoni, S., Bergenti, F., Mordonini, M., Adorni, G.: Evolving binary classifiers through parallel computation of multiple fitness cases. IEEE Transactions on Systems, Man, and Cybernetics, Part B 35(3), 548–555 (2005)
50. Cagnoni, S., Dobrzeniecki, A.B., Poli, R., Yanch, J.C.: Genetic algorithm-based interactive segmentation of 3D medical images. Image Vision Comput. 17(12), 881–895 (1999)
51. Calimeri, F., Ianni, G.: External sources of computation for answer set solvers. In: Baral, C., Greco, G., Leone, N., Terracina, G. (eds.) LPNMR 2005. LNCS (LNAI), vol. 3662, pp. 105–118. Springer, Heidelberg (2005)
52. Calimeri, F., Ianni, G.: Template programs for disjunctive logic programming: An operational semantics. AI Commun. 19(3), 193–206 (2006)
53. Calvanese, D., Giacomo, G.D., Lembo, D., Lenzerini, M., Rosati, R.: Tractable reasoning and efficient query answering in description logics: The *l-lite* family. Journal of Automated Reasoning 39(3), 385–429 (2007)
54. Calvanese, D., Guarino, N.: Ontologies and description logics. Intelligenza Artificiale 3(1-2), 21–27 (2006)
55. Campari, E., Levi, G.: A cellular automata model for highway traffic. European Physical Journal B 17(1), 159–166 (2000)
56. Cannata, N., Corradini, F., Merelli, E., Omicini, A., Ricci, A.: An agent-oriented conceptual framework for systems biology. In: T. Comp. Sys. Biology, pp. 105–122 (2005)
57. Carlucci Aiello, L., Levi, G.: The uses of metaknowledge in AI systems. In: Metalevel Architectures and Reflection, pp. 243–254. North-Holland, Amsterdam (1988)
58. Carlucci Aiello, L., Massacci, F.: Verifying security protocols as planning in logic programming. ACM Transactions on Computational Logic 2(4), 542–580 (2001)
59. Carpineto, C., Romano, G.: A lattice conceptual clustering system and its application to browsing retrieval. Machine Learning 24(2), 95–122 (1996)
60. Cascio, F., Console, L., Osella, M.G.M., Panati, A., Sottano, S., Dupré, D.T.: Generating on-board diagnostics of dynamic automotive systems based on qualitative models. AI Commuications 12(1-2), 33–43 (1999)

61. Castelfranchi, C.: Modelling social action for AI agents. Artificial Intelligence 103(1-2), 157–182 (1998)
62. Castellini, C., Giunchiglia, E., Tacchella, A.: SAT-based planning in complex domains: Concurrency, constraints and nondeterminism. Artificial Intelligence 147(1-2), 85–117 (2003)
63. Castelpietra, C., Guidotti, A., Iocchi, L., Nardi, D., Rosati, R.: Design and implementation of cognitive soccer robots. In: Birk, A., Coradeschi, S., Tadokoro, S. (eds.) RoboCup 2001. LNCS (LNAI), vol. 2377, pp. 312–318. Springer, Heidelberg (2002)
64. Ceci, M., Malerba, D.: Classifying web documents in a hierarchy of categories: a comprehensive study. Journal of Intelligent Information Systems 28(1), 37–78 (2007)
65. Cesa-Bianchi, N., Lugosi, G.: Potential-based algorithms in on-line prediction and game theory. Machine Learning 51(3), 239–261 (2003)
66. Cesta, A., Oddi, A., Smith, S.F.: A constraint-based method for project scheduling with time windows. Journal of Heuristics 8(1), 109–136 (2002)
67. Chella, A., Frixione, M., Gaglio, S.: Understanding dynamic scenes. Artif. Intell. 123(1-2), 89–132 (2000)
68. Chella, A., Gaglio, S., Pirrone, R.: Conceptual representations of actions for autonomous robots. Robotics and Autonomous Systems 34(4), 251–263 (2001)
69. Chella, A., Guarino, M.D., Infantino, I., Pirrone, R.: A vision system for symbolic interpretation of dynamic scenes using arsom. Applied Artificial Intelligence 15(8), 723–734 (2001)
70. Chittaro, L., Montanari, A.: Efficient temporal reasoning in the cached event calculus. Computational Intelligence 12, 359–382 (1996)
71. Chittaro, L., Ranon, R.: Hierarchical model-based diagnosis based on structural abstraction. Artificial Intelligence 155(1-2), 147–182 (2004)
72. Chittaro, L., Tasso, C., Toppano, E.: Putting functional knowledge on firmer ground. Applied Artificial Intelligence 8(2), 239–258 (1994)
73. Ciancarini, P., Omicini, A., Zambonelli, F.: Multiagent system engineering: The coordination viewpoint. In: Jennings, N.R. (ed.) ATAL 1999. LNCS, vol. 1757, pp. 250–259. Springer, Heidelberg (2000)
74. Cimatti, A., Pistore, M., Roveri, M., Traverso, P.: Weak, strong, and strong cyclic planning via symbolic model checking. Artificial Intelligence 147(1-2), 35–84 (2003)
75. Console, L., Dupré, D.T., Torasso, P.: A theory of diagnosis for incomplete causal models. In: IJCAI, pp. 1311–1317 (1989)
76. Console, L., Picardi, C., Dupré, D.T.: Temporal decision trees: Model-based diagnosis of dynamic systems on-board. Journal of Artificial Intelligence Research 19, 469–512 (2003)
77. Console, L., Picardi, C., Ribaudo, M.: Process algebras for systems diagnosis. Artificial Intelligence 142(1), 19–51 (2002)
78. Console, L., Torasso, P.: Automated diagnosis. Intelligenza Artificiale 3(1-2), 42–48 (2006)
79. Corno, F., Sánchez, E., Squillero, G.: Evolving assembly programs: how games help microprocessor validation. IEEE Transactions on Evolutionary Computation 9(6), 695–706 (2005)
80. Corradi, A., Omicini, A., Poggi, A. (eds.): WOA 2000: Dagli Oggetti agli Agenti. 1st AI*IA/TABOO Joint Workshop "From Objects to Agents": Evolutive Trends of Software Systems, Parma, Italy, 29-30 May 2000. Pitagora Editrice, Bologna (2000)
81. Costantini, S.: Meta-reasoning: A survey. In: Computational Logic: Logic Programming and Beyond, pp. 253–288 (2002)

82. Costantini, S.: Component-based answer set programming. In: Osorio, M., Provetti, A. (eds.) Latin-American Workshop on Non-Monotonic Reasoning, Proceedings of the 1st Intl. LA-NMR04 Workshop, Antiguo Colegio de San Ildefonso, Mexico City, D.F, Mexico, April 26, 2004. CEUR Workshop Proceedings, vol. 92, CEUR-WS.org (2004)
83. Costantini, S., Dell'Acqua, P., Tocchio, A.: Expressing preferences declaratively in logic-based agent languages. In: Proceedings of the 7th WOA 2006 Workshop, From Objects to Agents (Dagli Oggetti Agli Agenti), Catania, Italy, September 26-27, 2006. CEUR Workshop Proceedings, vol. 204, CEUR-WS.org (2006)
84. Costantini, S., Tocchio, A., Toni, F., Tsintza, P.: A multi-layered general agent model. In: Basili, R., Pazienza, M.T. (eds.) AI*IA 2007. LNCS (LNAI), vol. 4733, pp. 121–132. Springer, Heidelberg (2007)
85. Cucchiara, R., Perini, E., Pistoni, G.: Efficient stereo vision for obstacle detection and agv navigation. In: Cucchiara, R. (ed.) ICIAP, pp. 291–296. IEEE Computer Society Press, Los Alamitos (2007)
86. http://www.informatik.uni-trier.de/~ley/db/ (Viewed October 2007)
87. Delzanno, G., Gabbrielli, M.: Compositional verification of asynchronous processes via constraint solving. In: Caires, L., Italiano, G.F., Monteiro, L., Palamidessi, C., Yung, M. (eds.) ICALP 2005. LNCS, vol. 3580, pp. 1239–1250. Springer, Heidelberg (2005)
88. Dorigo, M., Stützle, T.: Ant Colony Optimization. MIT Press, Cambridge (2004)
89. Dovier, A., Formisano, A., Pontelli, E.: An experimental comparison of constraint logic programming and answer set programming. In: AAAI, pp. 1622–1625 (2007)
90. Dovier, A., Piazza, C., Rossi, G.: Multiset rewriting by multiset constraint solving. Romanian Journal of Information Science and Technology 4(1–2), 59–76 (2001)
91. Ernandes, M., Angelini, G., Gori, M.: Webcrow: A web-based system for crossword solving. In: AAAI, pp. 1412–1417 (2005)
92. Esposito, F., Giordana, A., Saitta, L.: Machine learning and data mining. Intelligenza Artificiale 3(1-2), 63–71 (2006)
93. Esposito, F., Malerba, D., Lisi, F.A.: Machine learning for intelligent processing of printed documents. Journal of Intelligent Information Systems 14(2-3), 175–198 (2000)
94. Esposito, F., Malerba, D., Semeraro, G.: A comparative analysis of methods for pruning decision trees. IEEE Transactions on Pattern Analysis and Machine Intelligence 19(5), 476–491 (1997)
95. Esposito, F., Semeraro, G., Fanizzi, N., Ferilli, S.: Multistrategy theory revision: Induction and abduction in INTHELEX. Machine Learning 38(1-2), 133–156 (2000)
96. Falcone, R., Castelfranchi, C.: The human in the loop of a delegated agent: the theory of adjustable social autonomy. IEEE Transactions on Systems, Man, and Cybernetics, Part A 31(5), 406–418 (2001)
97. Ferrari, G.L., Gnesi, S., Montanari, U., Pistore, M.: A model-checking verification environment for mobile processes. ACM Trans. Softw. Eng. Methodol. 12(4), 440–473 (2003)
98. Finzi, A., Pirri, F.: Combining probabilities, failures and safety in robot control. In: IJCAI, pp. 1331–1336 (2001)
99. Fioravanti, F., Pettorossi, A., Proietti, M.: Transformation Rules for Locally Stratified Constraint Logic Programs. In: Bruynooghe, M., Lau, K.-K. (eds.) Program Development in Computational Logic. LNCS, vol. 3049, pp. 291–339. Springer, Heidelberg (2004)
100. Fornara, N., Viganò, F., Colombetti, M.: Agent communication and artificial institutions. Autonomous Agents and Multi-Agent Systems 14(2), 121–142 (2007)

101. Franconi, E., Kifer, M., May, W. (eds.): ESWC 2007. LNCS, vol. 4519. Springer, Heidelberg (2007)

102. Frasconi, P., Gori, M., Sperduti, A.: A general framework for adaptive processing of data structures. IEEE Transactions on Neural Networks 9(5), 768–786 (1998)

103. Fumera, G., Roli, F.: A theoretical and experimental analysis of linear combiners for multiple classifier systems. IEEE Trans. Pattern Anal. Mach. Intell. 27(6), 942–956 (2005)

104. Galassi, U., Giordana, A., Saitta, L.: Incremental construction of structured hidden markov models. In: Veloso, M.M. (ed.) IJCAI 2007, Proceedings of the 20th International Joint Conference on Artificial Intelligence, Hyderabad, India, January 6-12, 2007, pp. 798–803 (2007)

105. Gandon, F., Poggi, A., Rimassa, G., Turci, P.: Multi-agent corporate memory management system. Applied Artificial Intelligence 16(9-10), 699–720 (2002)

106. Gangemi, A., Guarino, N., Masolo, C., Oltramari, A., Schneider, L.: Sweetening ontologies with DOLCE. In: Gómez-Pérez, A., Benjamins, V.R. (eds.) EKAW 2002. LNCS (LNAI), vol. 2473, pp. 166–181. Springer, Heidelberg (2002)

107. Gaspero, L.D., Schaerf, A.: A composite-neighborhood tabu search approach to the traveling tournament problem. Journal of Heuristics 13(2), 189–207 (2007)

108. Di Gaspero, L., di Tollo, G., Roli, A., Schaerf, A.: Hybrid local search for constrained financial portfolio selection problems. In: Van Hentenryck, P., Wolsey, L.A. (eds.) CPAIOR 2007. LNCS, vol. 4510, pp. 44–58. Springer, Heidelberg (2007)

109. Gerevini, A., Saetti, A., Serina, I.: Planning through stochastic local search and temporal action graphs in LPG. Journal of Artificial Intelligence Research 20, 239–290 (2003)

110. Ghilardi, S., Nicolini, E., Zucchelli, D.: A comprehensive framework for combined decision procedures. In: Gramlich, B. (ed.) FroCos 2005. LNCS (LNAI), vol. 3717, pp. 1–30. Springer, Heidelberg (2005)

111. Giacobini, M., Tomassini, M., Tettamanzi, A., Alba, E.: Selection intensity in cellular evolutionary algorithms for regular lattices. IEEE Transactions on Evolutionary Computation 9(5), 489–505 (2005)

112. Giacomo, G.D., Iocchi, L., Nardi, D., Rosati, R.: A theory and implementation of cognitive mobile robots. J. Log. Comput. 9(5), 759–785 (1999)

113. Giacomo, G.D., Lenzerini, M.: A uniform framework for concept definitions in description logics. CoRR cs.AI/9703101 (1997)

114. Giordana, A., Saitta, L.: Phase transitions in relational learning. Machine Learning 41(2), 217–251 (2000)

115. Giordano, L., Martelli, A., Schwind, C.: Specifying and verifying interaction protocols in a temporal action logic. J. Applied Logic 5(2), 214–234 (2007)

116. Giunchiglia, E., Narizzano, M., Tacchella, A.: Clause/term resolution and learning in the evaluation of quantified boolean formulas. Journal of Artificial Intelligence Research 26, 371–416 (2006)

117. Giunchiglia, F., Serafini, L.: Multilanguage hierarchical logics or: How we can do without modal logics. Artificial Intelligence 65(1), 29–70 (1994)

118. Gori, M., Marinai, S., Soda, G.: Artificial neural networks for document analysis and recognition. IEEE Transactions on Pattern Analysis and Machine Intelligence 27(1), 23–35 (2005)

119. Gori, M., Sperduti, A.: The loading problem for recursive neural networks. Neural Networks 18(8), 1064–1079 (2005)

120. Gregorio, S.D., Serra, R., Villani, M.: Applying cellular automata to complex environmental problems: The simulation of the bioremediation of contaminated soils. Theoretical Computer Science 217(1), 131–156 (1999)
121. Guarino, N., Welty, C.A.: Evaluating ontological decisions with ontoclean. Communications of ACM 45(2), 61–65 (2002)
122. Hoffmann, J., Bertoli, P., Pistore, M.: Web service composition as planning, revisited: In between background theories and initial state uncertainty. In: Proceedings of the Twenty-Second AAAI Conference on Artificial Intelligence, Vancouver, British Columbia, Canada, July 22-26, 2007, pp. 1013–1018. AAAI Press, Menlo Park (2007)
123. Intelligenza artificiale 3(1-2) (in English) (2006)
124. Iocchi, L., Nardi, D., Piaggio, M., Sgorbissa, A.: Distributed coordination in heterogeneous multi-robot systems. Auton. Robots 15(2), 155–168 (2003)
125. http://www.isaacrobot.it/ (Viewed October 2007)
126. Kakas, A.C., Mancarella, P.: On the relation between truth maintenance and abduction. In: Proc. of the first Pacific Rim International Conference on Artificial Intelligence, PRICAI-90, pp. 158–176 (1990)
127. Kimani, S., Lodi, S., Catarci, T., Santucci, G., Sartori, C.: VidaMine: a visual data mining environment. J. Vis. Lang. Comput. 1(15), 37–67 (2004)
128. Landwehr, N., Passerini, A., Raedt, L.D., Frasconi, P.: kFOIL: Learning simple relational kernels. In: AAAI (2006)
129. Lane, D., Serra, R., Villani, M., Ansaloni, L.: A theory-based dynamical model of innovation processes. ComPlexUs 2(3-4), 177–194 (2006)
130. Langdon, W.B., Poli, R.: Foundations of Genetic Programming, 2nd edn. Springer, Heidelberg (2005)
131. Leone, N., Pfeifer, G., Faber, W., Eiter, T., Gottlob, G., Perri, S., Scarcello, F.: The DLV system for knowledge representation and reasoning. ACM Transactions on Computational Logic 7(3), 499–562 (2006)
132. Licchelli, O., Esposito, F., Semeraro, G., Bordoni, L.: Personalization to improve searching in a digital library. In: Proceedings of the 3rd International Workshop on New Developments in Digital Libraries, NDDL, pp. 47–55 (2003)
133. Lisi, F.A.: A methodology for building semantic web mining systems. In: Esposito, F., Raś, Z.W., Malerba, D., Semeraro, G. (eds.) ISMIS 2006. LNCS (LNAI), vol. 4203, pp. 306–311. Springer, Heidelberg (2006)
134. Lisi, F.A., Malerba, D.: Inducing multi-level association rules from multiple relations. Machine Learning 55(2), 175–210 (2004)
135. Malerba, D., Appice, A., Ceci, M.: A data mining query language for knowledge discovery in a geographical information system. In: Database Support for Data Mining Applications, pp. 95–116 (2004)
136. Malerba, D., Esposito, F., Ceci, M., Appice, A.: Top-down induction of model trees with regression and splitting nodes. IEEE Transactions on Pattern Analysis and Machine Intelligence 26(5), 612–625 (2004)
137. Mamei, M., Zambonelli, F., Leonardi, L.: Co-fields: Towards a unifying approach to the engineering of swarm intelligent systems. In: Petta, P., Tolksdorf, R., Zambonelli, F. (eds.) ESAW 2002. LNCS (LNAI), vol. 2577, pp. 68–81. Springer, Heidelberg (2003)
138. Marconi, A., Pistore, M., Traverso, P.: Specifying data-flow requirements for the automated composition of web services. In: Fourth IEEE International Conference on Software Engineering and Formal Methods (SEFM 2006), Pune, India, 11-15 September 2006, pp. 147–156. IEEE Computer Society Press, Los Alamitos (2006)

139. Marriott, K., Stuckey, P.: Programming with constraints: an introduction. MIT Press, Cambridge (1998)
140. Marrocco, C., Molinara, M., Tortorella, F.: Exploiting auc for optimal linear combinations of dichotomizers. Pattern Recognition Letters 27(8), 900–907 (2006)
141. Cadoli, M., Donini, F.M., Liberatore, P., Schaerf, M.: Comparing space efficiency of propositional knowledge representation formalisms. In: KR, pp. 364–373 (1996)
142. Meisels, A., Schaerf, A.: Modelling and solving employee timetabling problems. Annals of Mathematics and Artificial Intelligence 39(1-2), 41–59 (2003)
143. Micheli, A., Sona, D., Sperduti, A.: Contextual processing of structured data by recursive cascade correlation. IEEE Transactions on Neural Networks 15(6), 1396–1410 (2004)
144. Milano, M. (ed.): Constraint and Integer Programming: Toward a Unified Methodology. Kluwer Academic Publishers, Dordrecht (2004)
145. Montanari, U.: Networks of constraints: Fundamental properties and applications to picture processing. Inf. Sci. 7, 95–132 (1974)
146. Nardi, D., Adorni, G., Bonarini, A., Chella, A., Clemente, G., Pagello, E., Piaggio, M.: ART99 - azzurra robot team. In: Veloso, M.M., Pagello, E., Kitano, H. (eds.) RoboCup 1999. LNCS (LNAI), vol. 1856, pp. 695–698. Springer, Heidelberg (2000)
147. Nicola, R.D., Ferrari, G.L., Pugliese, R.: KLAIM: A kernel language for agents interaction and mobility. IEEE Transactions on Software Engineering 24(5), 315–330 (1998)
148. Nolfi, S., Floreano, D.: Evolutionary Robotics. MIT Press, Cambridge (2000)
149. Olague, G., Cagnoni, S., Lutton, E.: Introduction to the special issue on evolutionary computer vision and image understanding. Pattern Recognition Letters 27(11), 1161–1163 (2006)
150. Olivetti, N.: Tableaux for nonmonotonic logics. In: D'Agostino, M., Gabbay, D., Haehnle, R., Posegga, J. (eds.) Hanbook of Tableaux Methods, Kluwer Academic Publishers, Dordrecht (1999)
151. Omicini, A., Zambonelli, F.: Coordination for internet application development. Autonomous Agents and Multi-Agent Systems 2(3), 251–269 (1999)
152. Palù, A.D., Dovier, A., Fogolari, F.: Constraint logic programming approach to protein structure prediction. BMC Bioinformatics 5, 186 (2004)
153. Palú, A.D., Dovier, A., Pontelli, E.: A constraint solver for discrete lattices, its parallelization, and application to protein structure prediction. Software Practice & Experience 37(13), 1405–1449 (2007)
154. Passerini, A., Frasconi, P., Raedt, L.D.: Kernels on prolog proof trees: Statistical learning in the ILP setting. Journal of Machine Learning Research 7, 307–342 (2006)
155. Perini, A., Susi, A.: Developing a decision support system for integrated production in agriculture. Environmental Modelling and Software 19(9), 821–829 (2004)
156. Pirri, F.: About implicit and explicit shape representation. In: Stock, O., Schaerf, M. (eds.) Reasoning, Action and Interaction in AI Theories and Systems. LNCS (LNAI), vol. 4155, pp. 141–158. Springer, Heidelberg (2006)
157. Piunti, M., Castelfranchi, C., Falcone, R.: Surprise as shortcut for anticipation: Clustering mental states in reasoning. In: Veloso, M.M. (ed.) IJCAI 2007, Proceedings of the 20th International Joint Conference on Artificial Intelligence, Hyderabad, India, January 6-12, 2007, pp. 507–512 (2007)
158. Policella, N., Wang, X., Smith, S., Oddi, A.: Exploiting temporal flexibility to obtain high quality schedules. In: Proc. AAAI-05 (2005)
159. Pollastri, G., Vullo, A., Frasconi, P., Baldi, P.: Modular DAG-RNN architectures for assembling coarse protein structures. Journal of Computational Biology 13(3), 631–650 (2006)

160. Portinale, L., Magro, D., Torasso, P.: Multi-modal diagnosis combining case-based and model-based reasoning: a formal and experimental analysis. Artificial Intelligence 158(2), 109–153 (2004)
161. Riguzzi, F.: ALLPAD: Approximate learning of logic programs with annotated disjunctions. Machine Learning (To appear 2008), http://dx.medra.org/10.1007/s10994-007-5032-8
162. http://robocare.istc.cnr.it/ (Viewed October 2007)
163. Rossi, F., Venable, B., Yorke-Smith, N.: Simple temporal problems with preferences and uncertainty. In: Proc. CP 2003 workshop on Online Constraint Solving: Handling Change and Uncertainty, Kinsale, Co. Cork, Ireland (2003)
164. Sadri, F., Toni, F., Torroni, P.: Dialogues for negotiation: agent varieties and dialogue sequences. In: Meyer, J.-J.C., Tambe, M. (eds.) ATAL 2001. LNCS (LNAI), vol. 2333, pp. 405–421. Springer, Heidelberg (2002)
165. Saitta, L., Zucker, J.D.: A model of abstraction in visual perception. Applied Artificial Intelligence 15(8), 761–776 (2001)
166. Serra, R., Villani, M., Colacci, A.: Differential equations and cellular automata models of the growth of cell cultures and transformation foci. Complex Systems 13(4), 347–380 (2001)
167. Serra, R., Villani, M., Semeria, A.: Genetic network models and statistical properties of gene expression data in knock-out experiments. Journal of Theoretical Biology 227(1), 149–157 (2004)
168. Serra, R., Zanarini, G.: Complex Systems and Cognitive Processes. Springer, Berlin (1990)
169. Shawe-Taylor, J., Cristianini, N.: Kernel Methods for Pattern Analysis. Cambridge University Press, Cambridge (2004)
170. Sperduti, A., Starita, A.: Supervised neural networks for the classification of structures. IEEE Transactions on Neural Networks 8(3), 714–735 (1997)
171. Stock, O., Strapparava, C.: Getting serious about the development of computational humor. In: IJCAI, pp. 59–64 (2003)
172. Stock, O., Zancanaro, M., Busetta, P., Callaway, C.B., Krüger, A., Kruppa, M., Kuflik, T., Not, E., Rocchi, C.: Adaptive, intelligent presentation of information for the museum visitor in peach. User Model. User-Adapt. Interact. 17(3), 257–304 (2007)
173. Sure, Y., Gómez-Pérez, A., Daelemans, W., Reinberger, M.L., Guarino, N., Noy, N.F.: Why evaluate ontology technologies? because it works? IEEE Intelligent Systems 19(4), 74–81 (2004)
174. http://www.swarm-bots.org/ (Viewed October 2007)
175. Tessaris, S., Franconi, E.: Rules and queries with ontologies: a unifying logical framework. In: Proceedings of the 2005 International Workshop on Description Logics (DL2005), July 26-28, 2005. CEUR Workshop Proceedings, vol. 147, CEUR-WS.org (2005)
176. Tettamanzi, A., Tomassini, M.: Soft Computing: Integrating Evolutionary, Neural and Fuzzy Systems. Springer, Berlin (2001)
177. Tomassini, M., Vanneschi, L., Collard, P., Clergue, M.: A study of fitness distance correlation as a difficulty measure in genetic programming. Evolutionary Computation 13(2), 213–239 (2005)
178. Torasso, P.: Multiple representations and multi-modal reasoning in medical diagnostic systems. Artificial Intelligence in Medicine 23(1), 49–69 (2001)
179. Toulminet, G., Bertozzi, M., Mousset, S., Bensrhair, A., Broggi, A.: Vehicle detection by means of stereo vision-based obstacles features extraction and monocular pattern analysis. IEEE Transactions on Image Processing 15(8), 2364–2375 (2006)

180. Trianni, V., Nolfi, S., Dorigo, M.: Cooperative hole avoidance in a *swarm-bot*. Robotics and Autonomous Systems 54(2), 97–103 (2006)
181. Veloso, M.M. (ed.): IJCAI 2007, Proceedings of the 20th International Joint Conference on Artificial Intelligence, Hyderabad, India, January 6-12 (2007)
182. Vérel, S., Collard, P., Tomassini, M., Vanneschi, L.: Fitness landscape of the cellular automata majority problem: View from the 'Olympus'. Theoretical Computer Science 378(1), 54–77 (2007)
183. Vullo, A., Frasconi, P.: Prediction of protein coarse contact maps. J. Bioinformatics and Computational Biology 1(2), 411–431 (2003)
184. Vullo, A., Frasconi, P.: Disulfide connectivity prediction using recursive neural networks and evolutionary information. Bioinformatics 20(5), 653–659 (2004)

# Intelligent User Profiling

Silvia Schiaffino[1,2] and Analía Amandi[1,2]

[1] ISISTAN Research Institute, Universidad Nacional del Centro de la Provincia de Buenos Aires, Campus Universitario, Argentina
[2] CONICET, Consejo Nacional de Investigaciones Científicas y Técnicas, Argentina
{sschia,amandi}@exa.unicen.edu.ar

**Abstract.** User profiles or user models are vital in many areas in which it is essential to obtain knowledge about users of software applications. Examples of these areas are intelligent agents, adaptive systems, intelligent tutoring systems, recommender systems, intelligent e-commerce applications, and knowledge management systems. In this chapter we study the main issues regarding user profiles from the perspectives of these research fields. We examine what information constitutes a user profile; how the user profile is represented; how the user profile is acquired and built; and how the profile information is used. We also discuss some challenges and future trends in the intelligent user profiling area.

## 1 Introduction

A profile is a description of someone containing the most important or interesting facts about him or her. In the context of users of software applications, a user profile or user model contains essential information about an individual user. The motivation of building user profiles is that users differ in their preferences, interests, background and goals when using software applications. Discovering these differences is vital to providing users with personalized services.

The content of a user profile varies from one application domain to another. For example, if we consider an online newspaper domain, the user profile contains the types of news (topics) the user likes to read, the types of news (topics) the user does not like to read, the newspapers he usually reads, and the user's reading habits and patterns. In a calendar management domain the user profile contains information about the dates and times when the user usually schedules each type of activity in which he is involved, the priorities each activity feature has for the user, the relevance of each user contact and the user's scheduling and rescheduling habits. In other domains personal information about the user, such as name, age, job, and hobbies might be important.

Not only the content of user profiles differs from one domain to another, but also how the information they contain is acquired. The content of a user profile can be explicitly provided by the user or it has to be learned using some intelligent

M. Bramer (Ed.): Artificial Intelligence, LNAI 5640, pp. 193–216, 2009.

technique. User profiling implies inferring unobservable information about users from observable information about them, that is, their actions or utterances (Zukerman and Albrecht, 2001). A wide variety of Artificial Intelligence techniques have been used for user profiling, such as case-based reasoning (Lenz et al, 1998; Godoy et al., 2004), Bayesian networks (Horvitz et al, 1998; Conati et al, 2002; Schiaffino and Amandi, 2005; Garcia et al, 2007), association rules (Adomavicius and Tuzhilin, 2001; Schiaffino and Amandi, 2006), genetic algorithms (Moukas, 1996; Yannibelli et al, 2006), neural networks (Yasdi, 1999; Villaverde et al, 2006), among others.

The purpose of obtaining user profiles is also different in the various areas that use them. In adaptive systems, the user profile is used to provide the adaptation effect, that is to behave differently for different users (Brusilovsky and Millán, 2007). In intelligent agents, particularly in interface agents, the user profile is used to provide personalized assistance to users with respect to some software application (Maes, 1994). In intelligent tutoring systems, the user profile or student model is used to guide students in their learning process according to their knowledge and learning styles (Garcia et al, 2007). In e-commerce applications the user or customer profile is used to make personalized offers and to suggest or recommend products the user is supposed to like (Adomavicius and Tuzhilin, 2001). In knowledge management systems, the skills a user or employee has, the roles he takes within an organization, and his performance in these roles are used by managers or project leaders to assign him to the job position that suits him best (Sure et al, 2000). In recommender systems the user profile contains ratings for items like movies, news or books, which are used to recommend potentially interesting items to him and to other users with similar tastes or interests (Resnick and Varian, 1997).

In this Chapter we study user profiles from the different perspectives mentioned above. In Section 2 we describe what information constitutes a user profile. In Section 3 we examine the different ways in which we can acquire information about a user and then build a user profile. Section 4 focuses on intelligent user profiling techniques. Finally, Section 5 presents some future trends.

## 2    User Profile Contents

A user profile is a representation of information about an individual user that is essential for the (intelligent) application we are considering. This section describes the most common contents of user profiles: user interests; the user's knowledge, background and skills; the user's goals; user behaviour; the user's interaction preferences; the user's individual characteristics; and the user's context. We analyze and provide examples for the different contents in areas like intelligent agents, adaptive systems, intelligent tutoring systems, recommender systems, and knowledge management systems.

## 2.1    Interests

User interests are one of the most important (and typically the only) part of the user profile in information retrieval and filtering systems, recommender systems, some interface agents, and adaptive systems that are information-driven such as encyclopedias, museum guides, and news systems (Brusilovsky and Millán, 2007). Interests can represent news topics, web page topics, document topics, work-related topics or hobbies-related topics. Sometimes user interests are classified as short-term interests or long-term interests. The interest of users in football may be a short-term interest if the user reads or listens to news about this topic only during the World Cup, or a long-term interest if the user is always interested in this topic. For example, *NewsDude* (Billsus and Pazzani, 1999), an interface agent that learns about a user's interests in daily news stories, considers information about recent events as short-term interests, and a user's general preferences for news stories as long-term interests.

The most common representation of user interests are keyword-based models. In these models interests are represented by weighted vectors of keywords. Weights traditionally represent the relevance of the word for the user or within the topic. These representations are common in the Information Filtering and Information Retrieval areas. For example *Letizia* (Lieberman et al, 2001a), a browsing assistant, uses TF-IDF (term frequency/inverse document frequency) vectors to model user interests. In this technique the weight of each word is calculated by comparing the word frequency in a document against the word frequency in all the documents in a corpus (Salton and McGill, 1983). This technique is also used in *NewsDude* (Billsus and Pazzani, 1999), where news stories are converted to TF-IDF vectors.

A more powerful representation of user interests is through topic hierarchies (Godoy et al, 2004). Each node in the hierarchy represents a topic of interest for a user, which is defined by a set of representative words. This representation technique is important when we want to model not only general user interests such as sports or economy, but also the sub-topics of these interests that are relevant to a given user. For example, the user profile can indicate that a certain user is interested in documents talking about a famous football player and not in sports or football in general. An example of a topic hierarchy containing a user's interests is shown in Figure 1.

Often, a topic ontology is used as the reference to construct a user interest profile. An ontology is a conceptualization of a domain into a human-understandable, but machine-readable format consisting of entities, attributes, relationships, and axioms (Guarino and Giaretta 1995). For instance, in *Quickstep* (Middleton et al, 2004), the authors represent user profiles in terms of a research paper topic ontology. This recommender system was built to help researchers in a computer science laboratory setting, representing user profiling with a research topic ontology and using ontological inference to assist the profiling process. Similarly, in (Liang et al, 2007) students' interests within an e-learning system are determined using a topic ontology.

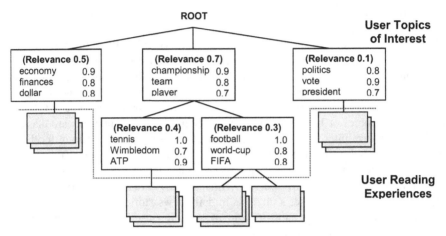

**Fig. 1.** Hierarchical representation of a user's interests

## 2.2    Knowledge, background and Skills

The knowledge the user has about the application domain, his background experience and his skills are important features within user profiles in different areas. In intelligent tutoring systems and adaptive educational systems, the student's knowledge about the subject taught is vital to provide proper assistance to the student or to adapt the content of courses according to it. This knowledge can be represented in different ways. The most common representation is through a model that keeps track of the student knowledge about every element in the course knowledge base. The idea is to mark each knowledge item X with a value calculated as "student knowledge of X". The value could be binary (knows - does not know), qualitative (good - average - bad) or quantitative, assigned as a probability of the student's familiarity with the item X. For instance, in Cumulate (Brusilovsky et al, 2005), the state of a student's knowledge is represented as a weighted overlay model covering a set of topics, and each educational activity can contribute to only one topic.

Another way of representing user's knowledge is through errors or misconceptions. In addition to (or instead of) modelling what the user knows, some works focus on modelling what the user does not know. For example, in (Chen and Hsieh 2005) the authors aim at diagnosing learners' common learning misconceptions during learning processes. They try to discover relationships between misconceptions.

Also, in many applications, the user's knowledge about the underlying domain is important. Some systems categorize users as expert, intermediate, or novice, depending on how well they know the application domain. For example, MetaDoc (Boyle and Encarnacion, 1994) considers the knowledge users have about Unix, which is the underlying application domain in this system.

Furthermore, user skills are key in areas like Knowledge Management. Within this area, skill management systems serve as technical platforms for mostly, though not exclusively, corporate-internal market places for skills and know-how. The systems are typically built on top of a database that contains profiles of employees and applicants. In this domain, profiles consist of numerous values for different skills and may be represented as vectors. In (Sure et al, 2000) authors use the integers "0" (no knowledge), "1" (beginner), "2" (intermediate) and "3" (expert) as skill values. Examples of skills can be "Programming in Y" or "Administration of Server X".

Finally, the user's background refers to those user's characteristics that are not directly related to the application domain. For instance, if we consider a tutoring system, the user's job or profession, his work experience, his traveling experience, the languages he speaks, among other information, constitute the user's background. As an application example, in (Cawsey et al, 2007) the authors describe an adaptive information system in the healthcare domain that considers users' literacy and medical background to provide them information that they can understand. The representation of users' background and skills is commonly done via stereotypes. We discuss them in Section 3.4.

## 2.3    Goals

Goals represent the user's objective or purpose with respect to the application he is working with, that is what the user wants to achieve. Goals are target tasks or subtasks at the focus of a user's attention (Horvitz et al, 1998). If the user is browsing the Web, his goal is obtaining relevant information (this type of goal is known as an information need). If the user is working with an e-learning system, his goal is learning a certain subject. In a calendar management system, the user's goals are scheduling new events or rescheduling conflicting events.

Determining what a user wants to do is not a trivial task. Plan recognition is a technique that aims at identifying the goal or intention of a user from the tasks he performs. In this context, a task corresponds to an action the user can perform in the software application, and a goal is a higher level intention of the user, which will be accomplished by carrying out a set of tasks. Systems using plan recognition observe the input tasks of a user and try to find all possible plans by which the observed tasks can be explained. These possible explanations or candidate plans are narrowed as the user continues performing further tasks. Plan recognition has been applied in different areas such as intelligent tutoring (Greer and Kohenn, 1995), interface agents (Lesh et al, 1999; Armentano and Amandi, 2006), and collaborative planning (Huber and Durfee, 1994).

Goals or intentions can be represented in different ways. Figure 2 shows a Bayesian network representation of a user's intentions in a calendar domain (Armentano and Amandi, 2006). In this representation, nodes represent user tasks and arcs represent probabilistic dependencies between tasks. Given evidence of a task performed by the user, the system can infer the next (most probable) task, and hence, the user's goal. Similarly, the *Lumiere* project at Microsoft Research (Horvitz et al., 1998) uses Bayesian networks to infer a user's needs by considering a user's background, actions and queries (help requests). Based on the beliefs of a user's needs and the utility theory of influence diagrams (an extension to Bayesian networks), an automated assistant provides help for users. In Andes (Gertner and VanLehn, 2000), plan recognition is necessary for the problem solving coach to select what step to suggest when a student asks for help. Since Andes wants to help students solve problems in their own way, it must determine what goal the student is probably trying to achieve, and suggest the action the student cannot perform due to lack of knowledge.

## 2.4    Behaviour

Usually, the user's behaviour with a software application is an important part of the user's profile. If a given user behaviour is repetitive, then it represents a pattern that can be used by an adaptive system or an intelligent agent to adapt a web site or to assist the user according to the behaviour learnt. The type of behaviour modelled depends on the application domain. For example, *CAP (Calendar APprentice)* learns the scheduling behaviour of its user and learns rules that enable it to suggest the meeting duration, location, time, and date (Mitchell et al, 1994). In an intelligent e-commerce system, a behavioural profile models the customer's actions (Adomavicius and Tuzhilin, 2001). Examples of behaviours in this domain are "When purchasing cereal, John Doe usually buys milk" and "On weekends, John Doe usually spends more than \$100 on groceries". In intelligent tutoring systems, the student behaviour is vital to assist him properly. In (Xu, 2002), a student profile is a set of $<t, e>$ pairs, where $e$ is a behaviour of the student and $t$ expresses the time when the behaviour occurs. $t$ could be a point in time or an interval of time. In this work, there are two main types of student behaviours, reading a particular topic and making a choice in a quiz.

Sometimes behaviours are routine, that is, they show some kind of regularity or seasonality. For example, QueryGuesser (Schiaffino and Amandi, 2005) models a user's routine queries to a database in a Laboratory Information Management System. In this agent, the user profile is composed of the queries each user performs

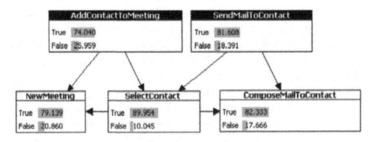

**Fig. 2.** Bayesian representation of a user's goals

and the moment when each query is generally made. The agent detects hourly, daily, weekly, and monthly behavioural patterns.

## 2.5    Interaction Preferences

A quite new component of a user profile is interaction preferences, that is, information about the user's interaction habits and preferences when he interacts with an interface agent (Schiaffino and Amandi, 2006). In interface agent technology, it is vital to know which agent's actions the user expects in different contexts and the modality of these actions. A user may prefer warnings, suggestions, or actions on the user's behalf. In addition, the agent can provide assistance by interrupting or not interrupting the user's work. A user interaction preference then expresses the preferred agent action and modality for different situations or contexts. As an illustration, consider an agent helping a user, John Smith, organize his calendar. Smith's current task is to schedule a meeting with several participants for the following Saturday in a free time slot. From past experience, the agent knows that one participant will disagree with the meeting date, because he never attends Saturday meetings. The agent can: warn the user about this problem, suggest another meeting date that considers all participant preferences and priorities, or do nothing. In this situation, some users would prefer a simple warning, while others would want suggestions about an alternative meeting date. In addition, when providing user assistance, agents can either interrupt the user's work or not. The agent must learn when the user prefers each modality. Information about these user preferences are kept in the user interaction profile, namely situations when the user: requires a suggestion to deal with a problem, needs only a warning about a problem, accepts an interruption from the agent, expects an action on his or her behalf, and wants a notification rather than an interruption.

## 2.6    Individual Characteristics

In some domains, personal information about the user is also part of the user profile. This item includes mainly demographic information such as gender, age, marital status, city, country, number of children, among other features. For example, Figure 3 shows the demographic profile of a customer in *Traveller*, a tourism recommender system that recommends package holidays and tours to customers.

On the other hand, a widely used user characteristic in intelligent tutoring systems and adaptive e-learning systems is the student's learning style. A learning-style model classifies students according to where they fit in a number of scales belonging to the ways in which they receive and process information. There have been proposed several models and frameworks for learning styles (Kolb 1984; Felder and Silverman, 1988; Honey and Mumford, 1992; Litzinger and Osif, 1993). For example, Felder and Silverman's model categorizes students as sensitive/intuitive, visual/verbal, active/reflective, and sequential/global, depending on how they learn. Various systems consider learning styles, such as ARTHUR (Gilbert

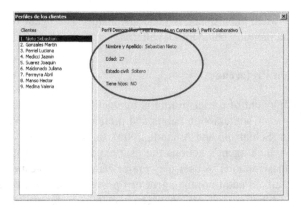

**Fig. 3.** Demographic profile of a customer in *Traveller*

and Han, 1999) which models three learning styles (visual-interactive, reading-listener, textual), CS388 (Carver et al, 1996) and MAS-PLANG (Peña et al., 2002) that use Felder and Silverman styles; the INSPIRE system (Grigoriadou et al., 2001) that uses the styles proposed by Honey and Mumford.

Finally, personality traits are also important features in a user profile. A trait is a temporally stable, cross-situational individual difference. One of the most famous personality models is OCEAN (Goldberg, 1993). This model comprises five personality dimensions: Openness to Experience, Conscientiousness, Extraversion, Agreeableness, and Neuroticism. Personality models and the methods to determine personality are subjects widely studied in psychology (McCrae and Costa, 1996; Wiggins et al, 1988). In the area of user profiling, various methods are used to detect user's personality. For example, in (Arya et al, 2006) facial actions are used as visual cues for detecting personality.

## 2.7    Contextual Information

The user's context is a quite new feature in user profiling. There are several definitions of context, mostly depending on the application domain. According to (Dey and Abwod, 1999), context is any information that can be used to characterize the situation of an entity. An entity is a person, place, or object that is considered relevant to the interaction between a user and an application, including the user and applications themselves. There are different types of contexts or contextual information that can be modelled within a user profile, as defined in (Goker and Myrhaug, 2002). The environmental context captures the entities that surround the user. These entities can, for instance, be things, services, temperature, light, humidity, noise, and persons. The personal context includes the physiological context and the mental context. The first part can contain information like pulse, blood pressure, weight, glucose level, retinal pattern, and hair colour. The latter part can contain information like mood, expertise, angriness, and stress. The social context describes the social aspects of the current user context. It can contain information about friends, neutrals, enemies, neighbours, co-workers, and relatives for instance. The spatio-temporal context describes aspects of the user context relating to the time and spatial extent for the user context. It can contain attributes like: time, location, or direction.

Context-aware systems (agents) are computing systems (agents) that provide relevant services and information to users based on their situational conditions or contexts (Dey and Abwod, 1999). In (Schiaffino and Amandi, 2006), for example, different types of assistance actions are executed by an agent depending on the task the user is carrying out and on the situation in which the user needs assistance. As regards users' emotions or mood, *RoCo* (Ahn and Picard, 2005) models different users' states, namely attentive, distracted, slumped, showing pleasure, showing displeasure, and acts accordingly. Other examples of context-aware systems based on the user location are various tourist guide projects where information is displayed depending on the current location of the user, such as (Yang et al, 1999).

## 2.8    Group Profiles

In contrast to individual user profiles, group profiles aim at combining individual user profiles to model a group. Group profiles are vital in those domains where it is necessary to make recommendations to groups of users rather than to individual users. Examples of these domains are tourism recommendation systems, movie recommenders, and adaptive television. In the first type of application, we find INTRIGUE (Ardissono et al, 2002), which recommends places to visit for tourist groups taking into account characteristics of subgroups within that group (such as children and disabled). Similarly, CATS (Collaborative Advisory Travel System) allows a group of users to simultaneously collaborate on choosing a skiing holiday package which satisfies the group as a whole (McCarthy et al, 2006). Group user feedback is used to suggest products that satisfy the individual and the group.

As regards TV, in (Masthoff, 2004) the authors discuss different strategies for combining individual user profiles to adapt to groups in an adaptive television application. In (Yu et al, 2006) the authors propose a recommendation scheme that merges individual user profiles to form a common user profile, and then generates common recommendations according to the common user profile.

## 3    Obtaining User Profiles

To build a user profile, the information needed can be obtained explicitly, that is provided directly by the user, or implicitly, through the observation of the user's actions. In this section we describe these alternatives.

## 3.1    Explicit Information

The simplest way of obtaining information about users is through the data they input via forms or other user interfaces provided for this purpose. Usually, this type of information is optional since users are not willing to fill in long forms providing information about them. Generally, the information gathered in this way is demographic, such as the user's age, gender, job, birthday, marital status, and hobbies. For eample, in (Adomavicius and Tuzhilin, 2001) this information consti-tutes the factual profile (name, gender, and date of birth), which is obtained by the e-commerce system from the customer's data.

In addition, personal interests can be informed explicitly. For example, in *NewsAgent* (Godoy et al, 2004) the user can indicate which sections of a digital newspaper he likes to read, which newspaper he prefers, or indicate general inter-esting topics, such as football, through a user interface, and he can also rate pages as interesting or uninteresting while he is reading. Figure 4 shows the user inter-faces for these purposes. In *Syskill & Webert* (Pazzani et al, 1996), users make explicit relevance judgments of pages explored while browsing the Web. *Syskill & Webert* learns a profile from the user's ratings of pages and uses this profile to suggest other pages. The user can rate a page as either hot (two thumbs up), luke-warm (one thumb up and one thumb down), or cold (two thumbs down). The *Apt Decision* agent (Shearin and Lieberman, 2001) learns user preferences in the do-main of rental real estate by observing the user's critique of apartment features. Users provide a small number of criteria in the initial interaction consisting of number of bedrooms, city, and price, then receive a display of sample apartments, and then react to any feature of any apartment independently, in any order.

Another way of providing explicit information is through the "Programming by Example" (PBE) or "Programming by Demonstration" paradigm (Lieberman, 2001b). In this approach, the user demonstrates examples to the computer. A

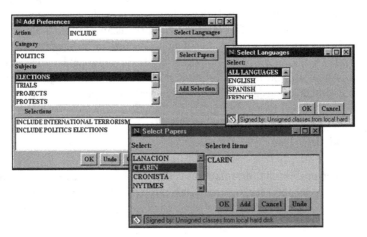

**Fig. 4.** Providing explicit information about a user's interests

software agent records the interactions between the user and a conventional interface, and writes a program that corresponds to the user's actions. The agent can then generalize the program so that it can work in other situations similar to, but not necessarily exactly the same as, the examples on which it is taught. For example, in (Ruvini and Dony, 2001) a software agent detects habitual patterns in a conventional programming language environment, Smalltalk, and automates those patterns.

## 3.2     Observation of a User's Actions

There are various problems with explicit user information. First, users are generally not willing to provide information by filling in long forms. Second, they not always tell or write the truth when completing forms about themselves. Third, although some of them might be willing to provide data, they sometimes do not know how to express their interests or what they really want. Thus, the most widely used method for obtaining information about users is observing their actions with the underlying application, recording or logging these actions, and discovering patterns from these logs through some Machine Learning or Data Mining technique.

In order to learn a user profile from a user's actions, there are certain conditions that must be fulfilled. The user behaviour has to be repetitive, that is the same actions have to be performed under similar conditions in different time points. If there is no repetition, no pattern can be discovered. In addition, the behaviour observed has to be different for different users. If not, there is no need for building an individual user profile.

For example, *PersonalSearcher* (Godoy et al, 2004) unobtrusively observes a user's browsing behaviour in order to approximate the degree of user interest in each visited web page. In order to accomplish this goal, for each read page in a standard browser the agent observes a set of implicit indicators in a process known as implicit feedback (Oard and Kim, 1998). Implicit interest indicators used by *Personal Searcher* include the time consumed in reading a web page (considering its length), the amount of scrolling in a page, and whether it was added to the list of bookmarks or not. Similarly, *NewsAgent* monitors users' behaviour while they are reading newspapers on the web and it records information about the different articles they read and some indicators about their relevance to the user.

A key characteristic of learning through observation is that of adapting to the user's changing interests, preferences, habits and goals. The user profiling techniques used have to be able to adapt the content of the user profile as new observations are recorded. User feedback plays a fundamental role in this task, as explained in the next section.

### 3.3     User Feedback

User feedback is a key source of learning in interface agent technology. This feedback may be explicit, when users explicitly evaluate an agent's actions through a user interface provided for that purpose, or implicit, when the agent observes a user's actions after assisting him to detect some implicit evaluation of its assistance. The explicit feedback can be simple or complex. It is simple when the user is required to evaluate the agent's assistance according to a quantitative or a qualitative scale (for example 0 to 10 or relevant or irrelevant) or to just press a dislike/like button. However, it becomes more complicated when the user is required to provide big amounts of information in various steps. Mostly, an interface agent has to learn from implicit feedback since the explicit feedback is not always available. As said before, the reason is that not all users are willing to provide explicit feedback, mainly if this demands of them a lot of time and effort. For example, *NewsDude* (Billsus and Pazzani, 1999) supports the following feedback options: interesting, not interesting, I already know this, tell me more, and explain. In *NewsAgent* (Godoy et al, 2004), once the agent shows the personalized newspaper to the user, he can rate each news item contained in the newspaper as interesting or uninteresting, as shown in Figure 5.

User ratings in recommender systems can be considered also as user feedback (or as explicit information as well). In these systems, users rate items such as movies they have seen or books they have read. These ratings are used both to build the user profile and to recommend potentially interesting items to other users similar to the user under consideration. This last type of recommendation is known as collaborative recommendation or collaborative filtering. For example, *MovieLens[1]* uses user ratings to generate personalized recommendations for other movies the user will like and dislike.

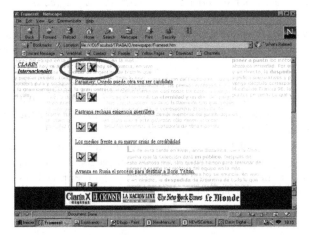

**Fig. 5.** Providing user feedback in *NewsAgent*

---

[1] http://movielens.umn.edu/

## 3.4    Stereotypes

A stereotype is the representation of relevant common characteristics of users pertaining to specific user subgroups of an application system (Kobsa, 2001). Stereotypes were the first attempt to differentiate a user from other users (Rich, 1979; Rich, 1989). Often, different system functionality was provided to users depending on their stereotype. The most popular stereotypes are: novice, intermediate or expert user. For example, UMT (Brajnik and Tasso, 1994) allows the user model developer the definition of hierarchically ordered user stereotypes; and BGP-MS (Kobsa and Pohl, 1995) allows assumptions about the user and stereotypical assumptions about user groups to be represented in a first-order predicate logic. Other examples of user stereotypes were presented in Section 2.2.

Stereotypes are useful when no other information about a user is available, that is, when the user has not used the system yet. This is the idea of um Toolkit (Kay, 1990), where information about a user stereotype is used as default information. Stereotypes enable the classification of users as belonging to one or more of a set of subgroups, and also the integration of the typical characteristics of these subgroups into the individual user profile.

# 4    Intelligent User Profiling Techniques

Intelligent user profiling implies the application of intelligent techniques, coming from the areas of Machine Learning, Data Mining or Information Retrieval, for example, to build user profiles. The data these techniques use to automatically build user profiles are obtained mainly from the observation of a user's actions, as described in the previous section. In this section we briefly describe three techniques widely used for user profiling and we present examples of their use that were developed in our research group.

## 4.1    Bayesian Networks

In the last decade, interest has been growing steadily in the application of Bayesian representations and inference methods for modelling the goals, preferences, and needs of users (Horvitz et al, 1998). A Bayesian network (BN) is a compact, expressive representation of uncertain relationships among variables of interest in a domain. A BN is a directed acyclic graph where nodes represent random variables and arcs represent probabilistic correlations between variables (Jensen, 2001). The absence of edges in a BN denotes statements of independence. A BN also represents a particular probability distribution, the joint distribution over all the variables represented by nodes in the graph. This distribution is specified by a set of conditional probability tables (CPT). Each node has an associated CPT that specifies the probability of each possible state of the node given each possible combination of states of its parents. For nodes without parents, probabilities are not conditioned on other nodes; these are called the prior or marginal probabilities of these variables.

The mathematical model underlying BN is Bayes' theorem, which is shown in Equation 1. Bayes' theorem relates conditional and marginal probabilities. It yields the conditional probability distribution of a random variable A, assuming we know: information about another variable B in terms of the conditional probability distribution of B given A, and the marginal probability distribution of A alone. Equation 1 reads: the probability of A given B equals the probability of B given A times the probability of A, divided by the probability of B.

An important characteristic of BN is that Bayesian inference mechanisms can be easily applied to them. The goal of inference is typically to find the conditional distribution of a subset of the variables, conditioned on known values for some other subset (the evidence). Thus, a BN can be considered a mechanism for automatically constructing extensions of Bayes' theorem to more complex problems.

$$P(A/B) = (P(B/A) P(A)) / P(B) \tag{1}$$

As an example of using BN for user profiling, consider the work presented in (Garcia et al, 2007), where BN are used to model a student's behaviour with an e-learning system and to detect his learning style. In this example, random variables represent the different dimensions of Felder's learning styles (Felder and Silverman, 1988) and the factors that determine each of these dimensions. The dimensions modelled are perception, processing and understanding. The values these variables can take are sensory/intuitive, active/reflective, and sequential/global respectively. The factors that determine them are extracted from the interactions between the student and a web-based education system. Thus, a BN models the relationships between the dimensions of learning styles and the factors determining them. A part of the BN proposed in this work is shown in Figure 6. This network models the relationships between the participation of a student in chats and forums and the processing style of this student. Thus, the BN has three nodes: chat, forum, and processing. The "chat" node has three possible states: participates, listens, no participation. The "forum" node has four possible states: replies messages, reads messages, posts messages and no participation. Finally, the "processing" node has two possible values, namely active and reflective.

The Bayesian model is completed with the simple probability tables for the independent nodes and the CPT for the dependent nodes. The values of the simple probabilities are obtained by analyzing a student's log file. The probability functions associated with the independent nodes are gradually obtained by observing the student interaction with the system. The values of the CPT are set by combining expertise knowledge and experimental results. As an example, Figure 6 shows the probability values obtained for a certain student for the "chat", "forum" and "processing" nodes. We can observe the marginal probabilities for the independent nodes, namely chat and forum, and the CPT for the processing node.

Once the BN is built, the student learning style is determined via Bayesian inference. The authors infer the values of the nodes corresponding to the dimensions

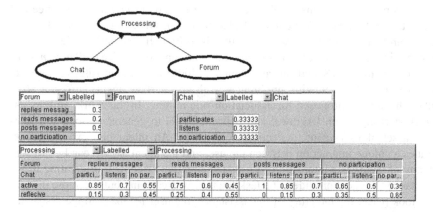

**Fig. 6.** Building a student profile with a BN

of a learning style given evidence of the student's behaviour with the system. The learning style of the student is the one having the greatest posterior probability value. In the simple example in Figure 6, given evidence of the utilization of the chat and forum facilities, we could infer whether the student processes information actively (discussing, in groups) or reflectively (by himself).

There are various works that use BN for user profiling. For example, the Lumiere project at Microsoft Research (Horvitz et al., 1998) uses BN to infer a user's needs by considering a user's background, actions and queries. Based on the beliefs of a user's needs an automated assistant provides help for users. In (Sanguesa et al, 1998) the authors use BN to model the profile of a web visitor and they use this profile to recommend interesting web pages. ANDES (Gertner and VanLehn, 2000) and SE-Coach (Conati and VanLehn, 2000) use this technique to model student knowledge in Physics. In (Gamboa and Fred, 2001) the authors use BN to assess students' state of knowledge and learning preferences in an intelligent tutoring system.

## 4.2    Association Rules

Association rules are a data mining technique widely used to discover patterns from data. They have also been used to learn user profiles in different areas, mainly in those related to e-commerce (Adomavicius and Tuzhilin, 2001) and web usage (Gery and Hadad, 2003). An association rule is a rule which implies certain association relationships among a set of objects in a given domain, such as they occur together or one implies the other. Association rule mining is commonly stated as follows (Agrawal and Srikant, 1994): Let $I$ be a set of items and $D$ be a set of transactions, each consisting of a subset $X$ of items in $I$. An association rule is an implication of the form X$\rightarrow$Y, where $X \subset I$, $Y \subset I$ and $X \cap Y = \varnothing$. $X$ is the antecedent of the rule and $Y$ is the consequent. The rule has support $s$ in $D$ if $s\%$ of the transactions in $D$ contains $X \cap Y$. The rule $X \rightarrow Y$ holds in $D$ with confidence $c$ if $c\%$ of transactions in $D$ that contain $X$ also contain $Y$. Given a transaction database $D$, the problem of mining association rules is to find all association rules that satisfy: minimum support (called *minsup*) and minimum confidence (called *minconf*).

For example, in (Schiaffino and Amandi, 2006) association rules are used to discover a user's interaction preferences with an interface agent. Different algorithms have been proposed for association rule mining. In this work, authors use the Apriori algorithm (Agrawal and Srikant, 1994) to generate association rules from a set of user-agent interaction experiences. An interaction experience describes a unique interaction between the user and the agent, which can be initiated by any of them. The interaction records the situation or context originating it, the assistance the agent provided, the task the user was carrying out when the interaction took place, the modality of the assistance, the user feedback to the assistance type and the modality (if available), and an evaluation of the interaction (success, failure, or undefined).

Association rule mining algorithms tend to produce huge amounts of rules, most of them irrelevant or uninteresting. Therefore, some post-processing steps are needed to obtain valuable information to build a user profile. In the example we are describing, the rules generated by Apriori are automatically post-processed in order to derive useful knowledge about the user from them. Post-processing includes detecting the most interesting rules, eliminating redundant and insignificant rules, eliminating contradictory weak rules, and summarizing the information in order to formulate hypotheses about a user's preferences more easily. For filtering rules, the authors use templates to express and select relevant rules. For example, they are interested in those association rules of the form "*situation, assistance action → user feedback, evaluation*"; and also in association rules of the form "*situation, modality, [user task], [relevance], [assistance action] → user feedback, evaluation*", where brackets mean that the attributes are optional. To eliminate redundant rules, they use a subset of the pruning rules proposed in (Shah, 1999). Basically, these pruning rules state that given the rules A,B→C and A→C, the first rule is redundant because it gives little extra information. Thus, it can be deleted if the two rules have similar confidence values. Similarly, given the rules A→B and A→B,C, the first rule is redundant since the second consequent is more specific. Thus, the redundant rule can be deleted provided that both rules have similar confidence values. A contradictory rule is one indicating a different assistance action (modality) for the same situation, and having a small confidence value with respect to the rule being compared. After pruning, rules are grouped by similarity and a hypothesis is generated considering: a main rule, positive evidence (redundant rules that could not be eliminated), and negative evidence (contradictory rules not eliminated). Once a hypothesis is formulated, the profiling algorithm computes the certainty degree of the hypothesis by taking into account the support values of the main rule, the positive and the negative evidence. The whole user profiling process is summarized in Figure 7.

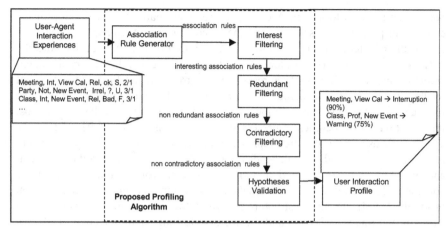

**Fig. 7.** User profiling with association rules

Other works have utilized association rules for building user profiles. In (Chen and Hsieh 2005) the authors use this technique for diagnosing learners' common learning misconceptions during learning processes. In this work, the association rules state that if misconception A occurs then misconception B is probable to occur. In (Adomavicius and Tuzhilin, 2001) association rules are used to build customers' profiles in an e-commerce application. Association rules indicate relationships among items bought by a customer.

## 4.3    Case-Based Reasoning

CBR is a technique that solves new problems by remembering previous similar experiences (Kolodner, 1993). A case-based reasoner represents problem-solving situations as cases. Given a new situation, it retrieves relevant cases (the ones matching the current problem) and it adapts their solutions to solve the problem. In an interpretative approach, CBR is applied to accomplish a classification task, that is, find the correct class for an unclassified case. The class of the most similar past case becomes the solution to the classification problem.

CBR has been used to build user profiles in areas like information retrieval and information filtering (Lenz et al, 1998; Smyth and Cotter, 1999). For example, in (Godoy et al, 2004) CBR is used to obtain a user interest profile. In this domain, cases represent relevant documents read by a user on the web. Each case records the main characteristics of a document, which enable the reasoner to determine its topic considering previous readings. Document topics represent a user's interests and they constitute case solutions. A topic or category is extensionally defined by a set of cases sharing the same solution. Using this approach, previously read documents can help to categorize new ones into specific categories, assuming that similar documents share the same topic.

Cases represent specific and contextual knowledge that describes a particular situation. In the example we are describing, cases represent readings of web documents relevant to a user. A case has three main parts: the description of the situation or problem, the solution, and the outcome or results of applying the solution to the problem. In this example, the description of the situation includes the URL (Uniform Resource Locator) of the document, a general category the document belongs to, a vector of representative words, and the time the user spent reading it. The solution is a code number that identifies a certain topic of interest for a user and the outcome describes in some way the user's feedback. The most important part of a document representation as a case is the list of relevant words. The authors use a bag-of-words representation where each word in the document has an associated weight that depends on the frequency of a word in the document and an additive factor defined in terms of several word characteristics in the document. The additive factor is calculated taking into account the word location inside the HTML document structure (words in the title are more important than words in the document body) and the word style (bold, italic or underlined). An example of a case in this domain is shown in Figure 8.

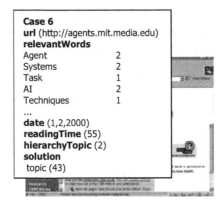

**Fig. 8.** Case representing an interesting web page

The comparison of cases is performed through a number of dimensions that describe them. A similarity function is defined for each of these dimensions, the most important being the one that measures the similarity between the relevant word lists. This similarity is calculated by the inner product with cosine normalization (Salton and McGill, 1983). A numerical evaluation function that combines the matching of each dimension with the importance value assigned to that dimension, is used to obtain the global similarity between the entry case $C_E$ and the retrieved one $C_R$. The function used is the formula in Equation 2, where $w_i$ is the importance of the dimension $i$ given by the agent designer; $sim_i$ is a similarity function for this dimension; $f_i^E$, $f_i^R$ are the values for the feature $f_i$ in both cases. If the similarity value obtained is higher than a given threshold, the cases are considered similar, and then, we can conclude that both cases are in the same user interest topic.

$$S(C_E, C_R) = \sum_{i=1}^{n} w_i * sim_i (f_i^E, f_i^R) \tag{2}$$

## 4.4    Other User Profiling Techniques

Many other Machine Learning techniques have been used for user profiling, such as genetic algorithms, neural networks, kNN-algorithm, clustering, and classification techniques such as decision trees or naïve Bayes classifier. For example, *Personal WebWatcher* (Mladenic, 1996) and *Syskill&Webert* (Pazzani et al, 1996) use naive Bayes classifiers for detecting users' interests when browsing the web. *Amalthaea* (Moukas, 1996) uses genetic algorithms to evolve a population of vectors representing a user's interests. The user profile is used to discover and filter information according to the user's interests. *NewsDude* (Billsus and Pazzani, 1999) obtains a short-term interest user profile using the k-NN algorithm and a long-term interest profile using a naïve Bayes classifier. *PersonalSearcher* (Godoy and Amandi, 2006) uses a clustering algorithm to categorize web documents and hence determine a user's interest profile. *SwiftFile* uses a TF-IDF style classifier to organize emails (Segal and Kephart, 2000). *CAP* uses decision trees to learn users' scheduling preferences (Mitchell et al., 1994).

Combinations of different techniques have also been used for building user profiles. For example, in (Martin-Bautista et al, 2000) the authors combine genetic algorithms and classification techniques (fuzzy logic) to build user profiles from a collection of documents previously retrieved by the user. In (Schiaffino and Amandi, 2000) case-based reasoning and Bayesian networks are combined to learn a user profile in a LIMS (Laboratory Information Management System). The user profile comprises routine user queries that represent a user's interests in the LIMS domain. In (Ko and Lee,2000) the authors combine genetic algorithms and a naive Bayes classifier to recommend interesting web documents to users .

## 5    Future Trends

We have studied in this Chapter the main issues concerning user profiles: how a user profile is composed; how a user profile can be acquired; and how a user profile can be used. We have seen that user profiles are vital in many areas; many of them in constant evolution and some new ones. Thus, researchers in the area of user profiling have to fulfill the expectations of these new trends and include new components as part of a profile and develop new techniques to build them.

As regards user profile contents, in recent years there has been increasing interest in modelling users' emotions and moods as part of the user profiles in areas such as social computing and intelligent agents. Emotional state has a similar structure to personality (described in Section 2.6), but it changes over time. The emotional state is a set of emotions that have a certain intensity. For example, the OCC model (Ortony et al, 1988) defines 22 emotions. An example of work in this direction is *AutoTutor* (D'Mello et al, 2007), which tries to determine students' emotions as they interact with an intelligent tutoring system. It uses several non-intrusive sensing devices to obtain this information. *AutoTutor* analyzes facial expressions, posture patterns, and conversational cues to determine a student's emotional state.

With respect to contextual information about a user, the developments in the areas of ubiquitous computing, mobile devices, and physical sensors enable the incorporation of new features in user profiles such as the focus of user attention (detected via eye-tracking), users' mood and emotions (detected analyzing facial expressions and body posture), temperature and humidity of the user's location, among others.

The area of Knowledge Management is acquiring great interest nowadays for organizations of different types. Within this area, user profiling is vital for different purposes. For example, building an employee profile focused on the employee's skills is important to place him in the position that best suits him. This is the purpose of skills management systems. Also, building a customer profile is important for customer relationship management (CRM). For example, in a credit card company, information such as the type of products the customer usually buys, how much he spends, when and where he buys what product, and how his family is composed, is key to offering him a personalized service.

Building group profiles is also a new tendency. Some works are being carried out in this direction (Jameson and Smyth, 2007). The challenges in this area are how to combine individual preferences into a group profile, how to help users to reach some kind of consensus, and how to make group recommendations trying to maximize average satisfaction, minimize misery and/or ensure some degree of fairness among participants.

# References

Adomavicius, G., Tuzhilin, A.: Using Data Mining Methods to Build Customer Profiles. IEEE Computer 34(2) (2001)

Agrawal, R., Srikant, R.: Fast Algorithms for Mining Association Rules. In: Proc. of the 20th Int'l Conference on Very Large Databases, Chile (1994)

Ahn, H., Picard, R.: Affective Cognitive Learning and Decision Making: A Motivational Reward Framework For Affective Agents. In: The 1st International Conference on Affective Computing and Intelligent Interaction, Beijing, China (2005)

Ardissono, L., Goy, A., Petrone, G., Segnan, M., Torasso, P.: Ubiquitous User Assistance in a Tourist Information Server. In: De Bra, P., Brusilovsky, P., Conejo, R. (eds.) AH 2002. LNCS, vol. 2347, pp. 14–23. Springer, Heidelberg (2002)

Armentano, M., Amandi, A.: A Bayesian Networks Approach to Plan Recognition for Interface Agents. In: Proc. Argentine Symposium on Artificial Intelligence, pp. 1–12 (2006)

Arya, A., Jefferies, N., Enns, J., DiPaola, S.: Facial actions as visual cues for personality. Computer Animation and Virtual Worlds 17(3-4), 371–382 (2006)

Baghaei, N., Mitrovic, A.: Evaluating a Collaborative Constraint-based Tutor for UML Class Diagrams. Frontiers in Artificial Intelligence and Applications, vol. 158. IOS Press, Amsterdam (2007)

Billsus, D., Pazzani, M.: A Personal News Agent that Talks, Learns and Explains. In: Proc. 3rd Int. Conf. on Autonomous Agents (Agents 99), Seattle, Washington (1999)

Boyle, C., Encarnacion, A.: MetaDoc: an adaptive hypertext reading system. User Modeling and User-adapted Interaction 4, 1–19 (1994)

Brajnik, G., Tasso, C.: A shell for developing non-monotonic user modeling systems. International Journal of Human-Computer Studies 40, 31–62 (1994)

Brusilovsky, P., Sosnovsky, S., Shcherbinina, O.: User Modeling in a Distributed E-learning Architecture. In: Ardissono, L., Brna, P., Mitrović, A. (eds.) UM 2005. LNCS (LNAI), vol. 3538, pp. 387–391. Springer, Heidelberg (2005)

Brusilovsky, P., Millán, E.: User Models for Adaptive Hypermedia and Adaptive Educational Systems. In: Brusilovsky, P., Kobsa, A., Nejdl, W. (eds.) Adaptive Web 2007. LNCS, vol. 4321, pp. 3–53. Springer, Heidelberg (2007)

Carver, C.A., Howard, R.A., Lavelle, E.: Enhancing student learning by incorporating learning styles into adaptive hypermedia. In: Proceedings of 1996 ED-MEDIA World Conf. on Educational Multimedia and Hypermedia, Boston, USA, pp. 118–123 (1996)

Cawsey, A., Grasso, F., Paris, C.: Adaptive Information for Consumers of Healthcare. In: Brusilovsky, P., Kobsa, A., Nejdl, W. (eds.) Adaptive Web 2007. LNCS, vol. 4321, pp. 465–484. Springer, Heidelberg (2007)

Chen, C., Hsieh, Y.: Mining Learner Profile Utilizing Association Rule for Common Learning Misconception Diagnosis. In: ICALT 2005, pp. 588–592 (2005)

Conati, C., VanLehn, K.: Toward computer-based support of meta-cognitive skills: A computational framework to coach self-explanation. The International Journal of Artificial Intelligence in Education 11, 389–415 (2000)

Conati, C., Gertner, A., VanLehn, K.: Using Bayesian Networks to Manage Uncertainty in Student Modeling. User Modeling and User-Adapted Interaction 12(4), 371–417 (2002)

Dey, A., Abwod, G.: Towards a better understanding of context and context-awareness. GVU Technical Report GIT-GVU-99-22 (1999), Also In the Workshop on The What, Who, Where, When, and How of Context-Awareness, CHI 2000

D'Mello, S.K., Picard, R.W., Graesser, A.C.: Towards an Affect-Sensitive AutoTutor. IEEE Intelligent Systems (Special issue on Intelligent Educational Systems) 22(4), 53–61 (2007)

Felder, R., Silverman, L.: Learning and Teaching Styles in Engineering Education. Engineering Education 78(7), 674–681 (1988)

Gamboa, H., Fred, A.: Designing intelligent tutoring systems: a bayesian approach. In: ICEIS Artificial Intelligence and Decision Support Systems, pp. 452–458 (2001)

Garcia, P., Amandi, A., Schiaffino, S., Campo, M.: Evaluating Bayesian Networks' Precision for Detecting Students' Learning Styles. Computers and Education 49(3), 794–808 (2007)

Gertner, A.S., VanLehn, K.: Andes: A Coached Problem Solving Environment for Physics. In: Gauthier, G., VanLehn, K., Frasson, C. (eds.) ITS 2000. LNCS, vol. 1839, pp. 133–148. Springer, Heidelberg (2000)

Gery, M., Hadad, H.: Evaluation of web usage mining approaches for user's next request prediction. In: Proceedings of the 5th ACM international workshop on Web information and data management, pp. 74–81 (2003)

Gilbert, J., Han, C.: Arthur: An Adaptive Instruction System Based on Learning Styles. In: Proceedings of International Conference on Mathematics / Science Education and Technology, pp. 100–105 (1999)

Godoy, D., Schiaffino, S., Amandi, A.: Interface Agents Personalizing Web-based Tasks. Cognitive Systems Research Journal (Special Issue on Intelligent Agents and Data Mining for Cognitive Systems) 5, 207–222 (2004)

Godoy, D., Amandi, A.: A Conceptual Clustering Approach for User Profiling in Personal Information Agents. AI Communications 19(3), 207–227 (2006)

Goker, A., Myrhaug, H.I.: User context and personalization. In: Proceedings of ECCBR Workshop on Case Based Reasoning and Personalization, UK (2002)

Goldberg, L.R.: The structure of phenotypic personality traits. American Psychologist 48, 26–34 (1993)

Greer, J., Koehn, G.: The peculiarities of plan recognition for intelligent tutoring systems. In: Proceedings of the workshop on The Next Generation of Plan Recognition Systems: Challenges for and Insight from Related Areas of AI, pp. 54–59 (1995)

Grigoriadou, M., Papanikolaou, K., Kornilakis, H., Magoulas, G.: INSPIRE: an intelligent system for personalized instruction in a remote environment. In: Proceedings of 3rd Workshop on Adaptive Hypertext and Hypermedia, pp. 13–24 (2001)

Guarnino, N., Giaretta, P.: Ontologies and knowledge bases: Towards a terminological clarification. In: Towards Very Large Knowledge Bases: Knowledge Building and Knowledge Sharing, pp. 25–32. IOS Press, Amsterdam (1995)

Honey, P., Mumford, A.: The Manual of Learning Styles. Maidenhead (1992)

Horvitz, E., Breese, J., Heckerman, D., Hovel, D., Rommelse, K.: The Lumiere project: Bayesian user modeling for inferring the goals and needs of software users. In: Proceedings of the 14th Conference on Uncertainty in Artificial Intelligence, pp. 256–265 (1998)

Huber, M., Durfee, E., Wellman, M.: The automated mapping of plans for plan recognition. In: Workshop on Distributed Artificial Intelligence, pp. 137–152 (1994)

Jameson, A., Smyth, B.: Recommendation to groups. In: Brusilovsky, P., Kobsa, A., Nejdl, W. (eds.) Adaptive Web 2007. LNCS, vol. 4321, pp. 596–627. Springer, Heidelberg (2007)

Jensen, F.: Bayesian Networks and Decision Graphs. Springer, Heidelberg (2001)

Kay, J.: um: a user modeling toolkit. In: Proc. 2nd International User Modeling Workshop, Hawaii, p. 11 (1990)

Ko, S.J., Lee, J.H.: Discovery of User Preference through Genetic Algorithm and Bayesian Categorization for Recommendation. In: Arisawa, H., Kambayashi, Y., Kumar, V., Mayr, H.C., Hunt, I. (eds.) ER Workshops 2001. LNCS, vol. 2465, pp. 471–484. Springer, Heidelberg (2002)

Kobsa, A., Pohl, W.: The BGP-MS user modeling system. User Modeling and User-Adapted Interaction 4(2), 59–106 (1995)

Kobsa, A.: Generic User Modeling Systems. User Modeling and User Adapted Interaction 11, 49–63 (2001)

Kolb, D.A.: Experiential learning: Experience as the source of learning and development. Prentice Hall, Upper Saddle River (1984)

Kolodner, J.: Case-based reasoning. Morgan Kaufmann, San Francisco (1993)

Lenz, M., Hubner, A., Kunze, M.: Question Answering with Textual CBR. In: Proceedings of the International Conference on Flexible Query Answering Systems, Denmark (1998)

Lesh, N.B., Rich, C., Sidner, C.L.: Using Plan Recognition in Human-Computer Collaboration. In: International Conference on User Modeling, June 1999, pp. 23–32 (1999)

Liang, Y., Zhao, Z., Zeng, Q.: Mining Users' Interests from Reading Behaviour in E-learning Systems. In: 8th ACIS International Conference on Software Engineering, Artificial Intelligence, Networking, and Parallel/Distributed Computing, pp. 417–422 (2007)

Lieberman, H., Fry, C., Weitzman, L.: Exploring the Web with Reconnaissance Agents. Communications of the ACM, 69–75 (Aug. 2001a)

Lieberman, H. (ed.): Your wish is my command: Programming by Example. Morgan Kaufman, San Francisco (2001b)

Litzinger, M.E., Osif, B.: Accommodating diverse learning styles: Designing instruction for electronic information sources. In: Shirato, L. (ed.) What is GoodInstruction Now? Library Instruction for the 90s, Pierian Press, Ann Arbor (1993)

Maes, P.: Agents that reduce work and information overload. Communications of the ACM 37(7), 31–40 (1994)

Martin-Bautista, M.J., Vila, M.A., Larsen, H.L.: Building adaptive user profiles by a genetic fuzzy classifier with feature selection. In: The Ninth IEEE International Conference on Fuzzy Systems (2000)

Masthoff, J.: Group Modeling: Selecting a Sequence of Television Items to Suit a Group of Viewers. User Modeling and User Adapted Interaction 14, 35–87 (2004)

McCarthy, K., Salamó, M., Coyle, L., McGinty, L., Smyth, B., Nixon, P.: Group Recommender Systems: A critiquing based approach. In: Proc. Intelligent User Interfaces, IUI 06 (2006)

McCrae, R., Costa Jr., P.T.: Toward a new generation of personality theories: Theoretical contexts for the five-factor model. In: Wiggins, J.S. (ed.) The five-factor model of personality: Theoretical perspectives, pp. 51–87. Guilford, New York (1996)

Middleton, S.E., Shadbolt, N.R., Roure, D.C.: Ontological user profiling in recommender systems. ACM Transactions on Information Systems (TOIS) 22(1), 54–88 (2004)

Mitchell, T., Caruana, R., Freitag, D., McDermoot, J., Zabowski, D.: Experience with a learning Personal Assistant. Communications of the ACM 37(7), 81–91 (1994)

Mladenic, D.: Personal WebWatcher: Implementation and Design. Technical Report IJS-DP-7472, Department of Intelligent Systems, J. Stefan Institute, Slovenia (1996)

Moukas, A.: Amalthaea: Information Discovery and Filtering using a Multi-agent Evolving Ecosystem. In: Proceedings of the Conference on the Practical Application of Intelligent Agents and MultiAgent Technology, London, UK (1996)

Oard, D., Kim, J.: Implicit feedback for recommender systems. In: Proceedings of the AAAI Workshop on Recommender Systems (1998)

Ortony, A., Clore, G.L., Collins, A.: The Cognitive Structure of Emotions. Cambridge University Press, Cambridge (1988)

Pazzani, M., Muramatsu, J., Billsus, D.: Syskill & Webert: Identifying Interesting Web Sites. AAAI/IAAI, vol. 1, pp. 54–61 (1996)

Peña, C., Marzo, J., de la Rosa, J.: Intelligent agents in a teaching and learning environment on the Web. In: Proceedings ICALT 2002, Rusia (2002)

Resnick, P., Varian, H.: Recommender Systems. Communications of the ACM 40(3), 56–58 (1997)

Rich, E.: User modeling via stereotypes. Cognitive Science 3, 355–366 (1979)

Rich, E.: Stereotypes and user modeling. In: Kobsa, A., Wahlster, W. (eds.) User Models in Dialog Systems, pp. 35–51. Springer, Heidelberg (1989)

Ruvini, J.D., Dony, C.: Learning Users' Habits to Automate Repetitive Tasks. In: Your wish is my command: Programming by Example, Morgan Kaufman, San Francisco (2001)

Salton, G.: Introduction to Modern Information Retrieval. McGraw-Hill, New York (1983)

Sanguesa, R., Cortés, U., Nicolás, M.: BayesProfile: application of Bayesian Networks to website user tracking. Technical Report, Universidad de Catalonia (1998)

Segal, R., Kephart, J.: Swiftfile: An intelligent assistant for organizing e-mail. In: AAAI 2000 Spring Symposium on Adaptive User Interfaces, Stanford, CA (2000)

Schiaffino, S., Amandi, A.: User Profiling with Case-Based Reasoning and Bayesian Networks. In: Open Discussion Proceedings, IBERAMIA-SBIA 2000, Atibaia, Brazil, pp. 12–21 (2000)

Schiaffino, S., Amandi, A.: An Interface Agent Approach to Personalize Users' Interaction with Databases. Journal of Intelligent Information Systems 25(3), 251–273 (2005)

Schiaffino, S., Amandi, A.: Polite Personal Agents. IEEE Intelligent Systems 21(1), 12–19 (2006)

Shah, D., Lakshmanan, L.V.S., Ramamritham, K., Sudarshan, S.: Interestingness and Pruning of Mined Patterns. In: Proceedings of the 1999 ACM SIGMOD Workshop on Research Issues in Data Mining and Knowledge Discovery (DMKD), Philadelphia (1999)

Shearin, S., Lieberman, H.: Intelligent Profiling by Example. In: Proceedings of the International Conference on Intelligent User Interfaces (IUI 2001), pp. 145–152 (2001)

Smyth, B., Cotter, P.: Surfing the Digital Wave: Generating Personalised TV Listings Using Collaborative, Case-Based Recommendation. In: Althoff, K.-D., Bergmann, R., Branting, L.K. (eds.) ICCBR 1999. LNCS (LNAI), vol. 1650, p. 561. Springer, Heidelberg (1999)

Sure, Y., Maedche, A., Staab, S.: Leveraging corporate skill knowledge - From ProPer to OntoProper. In: Proc. 3rd International Conf. on Practical Aspects of Knowledge Management, Basel, Switzerland (2000)

Villaverde, J., Godoy, D., Amandi, A.: Learning Styles Recognition in e-learning Environments using Feed-Forward Neural Networks. Journal of Computer Assisted Learning 22(3), 197–206 (2006)

Wiggins, J.S.: Personality and prediction: Principles of personality assessment. Krieger Publishing, Malabar (1988)

Xu, D., Wang, H., Su, K.: Intelligent Student Profiling with Fuzzy Models. In: Proceedings of the 35th Hawaii International Conference on System Sciences (2002)

Yang, J., Yang, W., Denecke, M., Waibel, A.: Smart sight: a tourist assistant system. In: Proceedings of Third International Symposium on Wearable Computers, p. 7378 (1999)

Yannibelli, V., Godoy, D., Amandi, A.: A Genetic Algorithm Approach to Recognize Students' Learning Styles. Interactive Learning Environments 14(1), 55–78 (2006)

Yasdi, R.: Learning User Model by Neural Networks. In: Proceedings of ICONIP '99, 6th International Conference on Neural Information Processing, pp. 48–53 (1999)

Yu, Z., Zhou, X., Hao, Y., Gu, J.: TV Program Recommendation for Multiple Viewers Based on user Profile Merging. User Modeling and User-Adapted Interaction 16(1), 63–82 (2006)

Zukerman, I., Albrecht, D.: Predictive Statistical Models for User Modeling. User Modeling and User-Adapted Interaction 11(1-2), 5–18 (2001)

# Supply Chain Business Intelligence: Technologies, Issues and Trends

Nenad Stefanovic[1] and Dusan Stefanovic[2]

[1] Zastava Automobiles, Information Systems Department, Kragujevac, Serbia
nenad@automobili.zastava.net, www.zastava-automobili.com
[2] Faculty of Science, University of Kragujevac, Serbia
dusans@kg.ac.yu, www.pmf.kg.ac.yu

**Abstract.** Supply chains are complex systems with silos of information that are very difficult to integrate and analyze. The best way to effectively analyze these disparate systems is the use of Business Intelligence (BI). The ability to make and then to process the right decision at the right time in collaboration with the right partners is the definition of the successful use of BI. This chapter discusses the need for Supply Chain Business Intelligence, introduces driving forces for its adoption and describes the supply chain BI architecture. The global supply chain performance measurement system based on the process reference model is described. The main cutting-edge technologies such as service-oriented architecture (SOA), business activity monitoring (BAM), web portals, data mining, and their role in BI systems are also discussed. Finally, key BI trends and technologies that will influence future systems are described.

## 1    Introduction – Supply Chain

Competing in today's business environment precipitates the need for successful integration and collaboration strategies among supply chain partners. The global environment is influenced by increased globalization and outsourcing, mergers, new technologies, and e-business. Shorter time-to-market, reduced product lifecycle, built-to-order strategies, pull systems and uncertainty force organizations to adopt new ways of doing business.

There was a lot of pressure on companies to increase profit, decrease cycle times, reduce inventories, improve service and adapt to forthcoming changes. Supply chain management (SCM) as a new management philosophy followed. Supply Chain Management was seen as a tool for gaining competitive advantage through real-time collaboration with trading partners, and offered a new way to rapidly plan, organize, manage, measure and deliver new products or services.

Many companies are beginning their search for a solution to implementing an electronically oriented supply chain management system that provides connection to customers and to suppliers. This integrated supply chain may be based on new

M. Bramer (Ed.): Artificial Intelligence, LNAI 5640, pp. 217–245, 2009.
© IFIP International Federation for Information Processing 2009

software solutions or based on enhanced communication capabilities. The ultimate objective is to create a "seamless system interface" that provides the capability to review and analyze varying elements of information. The objectives for analysis of this information are to create a more efficient supply chain characterized by [1]:

- Increased customer service levels;
- Decreased transaction costs;
- More efficient inventory investments;
- Reduced expenses for manufacturing;
- Increased responsiveness to customer demands;
- The ability to fulfill customer requirements more profitably;
- The ability to deliver high quality products in the shortest time;
- The ability to deliver products at the lowest cost;
- The ability to penetrate smaller, fragmented markets cost effectively;
- Greater linkages with key suppliers;
- Demand driven logistics;
- Capacity planning across the supply chain;
- Sharing of information with key suppliers thus reducing supplier costs.

In today's fast-changing global market organizations need to compete as supply chains, not as single business entities [2]. Additionally, an organization can participate in many supply chains, thus creating a complex supply network of interconnected processes (see Figure 1).

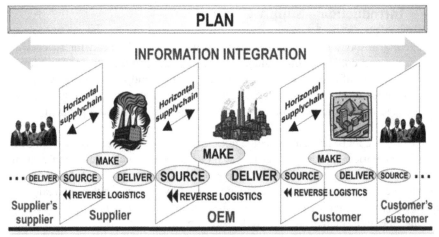

**Fig. 1.** The Supply Chain

Pushed by globalization and competitive forces, the classic linear supply chain has evolved into a complex value network of partners participating in a common business (Figure 2). These networks are expanding to include additional services provided by an increasing number of partners: customers, the government, financial services organizations, and so forth. Investments in systems or applications need to take into account the requirements or opportunities enabled by this increasing interconnectedness.

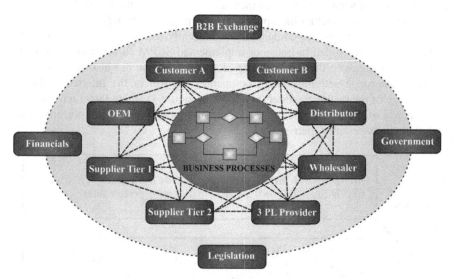

**Fig. 2.** The Supply Neitwork

However, despite the huge investment in SCM software systems, they did not provide desired Return On Investment (ROI). The main reason is that these systems mostly provide only transaction-based functionality. They lack the sophisticated analytical capabilities required to provide an integrated view of the supply chain. This is where Business Intelligence (BI) tools like data warehousing, ETL (Extraction, Transformation, and Loading), data mining, and OLAP (On-Line Analytical Processing) can help adequately analyze operational effectiveness across the supply chain.

## 1.1   Supply Chain Performance Measurement

Supply Chain Performance Measurement (SCPM) is vital for a company to survive in today's competitive business environment. Performance measurement is one of the key aspects of management. If a company does not have a clear understanding of how well its supply chains are performing, it will be very hard to manage them successfully.

Until a few years ago, there were several reasons why most companies did not implement supply chain performance measurement systems [3]:

1. No clear established approach or set of measures was available
2. Software vendor products offered only a limited range of supply chain metrics
3. Companies were too busy with other more important initiatives.

Measurement is important, as it affects behaviour that impacts supply chain performance [4]. As such, performance measurement provides the means by which a company can assess whether its supply chain has improved or degraded.

Only a few leading-edge companies are currently using true extended SCPM systems (either developed in-house or implemented SCPM software applications) that not only measure the performance of their enterprise but also that of their extended enterprise activities. Most companies are still in the Internal or Integrated stage of the maturity model (as shown in Figure 3) where they focus on the performance of their own enterprise and measure their supply chain performance with financially-oriented metrics.

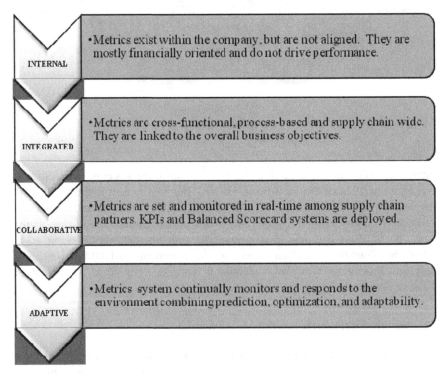

**Fig. 3.** A Supply Chain Performance Measurement Maturity Model

In response to some of these deficiencies in traditional accounting methods for measuring supply chain performance, a variety of measurement approaches have been developed, including the following: The Balanced Scorecard, The Supply Chain Council's SCOR Model, The Logistics Scoreboard, Activity-Based Costing (ABC), Economic Value Analysis (EVA), etc.

## 1.2    SCOR Model

SCOR model represents a universal approach to supply chain management that can be applied in diverse business domains [5]. SCOR combines business process engineering, benchmarking and best practices into a single framework. This standardization movement facilitates the deliverance of business content for the supply chain simulation model.

A standardized operations reference model provides significant benefits. Standardized models provide companies' maps toward business process engineering, establish benchmarking for performance comparison and uncover best business practices for gaining a competitive advantage. By standardizing supply chain operations and metrics for managing such operations, companies cannot only compare their results against others, but they are able to gain visibility of operations over a supply chain that may cross corporate borders. Partners in a supply chain can communicate more unambiguously and can collaboratively measure, manage and control their processes. Greater visibility over complicated orchestration of supply chain activities lets you fine-tune targeted problem areas and identify the cause-and-effect relationships in a supply chain network.

The process reference model contains:

- Standard descriptions of management processes
- A framework of relationships among the standard processes
- Standard metrics to measure process performance
- Management practices that produce best-in-class performance

According to SCOR, supply chain management consists of the following integrated processes: Plan (P), Source (S), Make (M), Deliver (D) and Return (R) — from the suppliers' supplier to the customers' customer, and all aligned with a company's operational strategy, material, work and information flows [6].

Additionally, SCOR includes a series of Enable (E) elements for each of the processes. Enable elements focus on information policy and relationships to enable the planning and execution of supply chain activities.

SCOR contains three levels of process detail. The top level consists of the aforementioned processes (plan, source, make, deliver and return) further subdivided into process types of planning, execution and enable (Figure 4).

The second SCOR level defines the configuration of planning and execution processes in material flow, using standard process categories such as make-to-order, engineer-to-order, or make-to-stock categories.

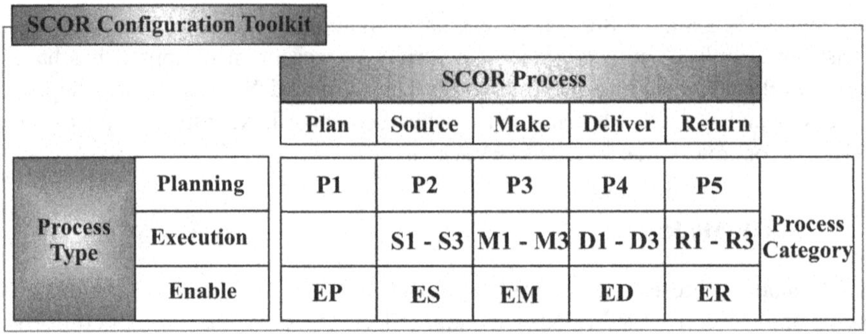

| SCOR Configuration Toolkit | | | | | | |
|---|---|---|---|---|---|---|
| | | **SCOR Process** | | | | |
| | | Plan | Source | Make | Deliver | Return |
| **Process Type** | **Planning** | P1 | P2 | P3 | P4 | P5 | Process Category |
| | **Execution** | | S1 - S3 | M1 - M3 | D1 - D3 | R1 - R3 | Process Category |
| | **Enable** | EP | ES | EM | ED | ER | |

**Fig. 4.** SCOR Process Categories

The third level consists of the actual process elements and their process flows. This bottom level of the SCOR model is most significant to analytics. It consists of process element definitions, inputs and outputs, relationships, performance metrics and best practices where applicable.

Metrics enable measurement and benchmarking of supply-chain performance. All process metrics are aspects of a performance attribute. The performance attributes for any given process are characterized as either customer-facing (reliability, responsiveness and flexibility) or internal-facing (cost and assets) metrics as shown in Table 1. Level 1 Metrics are primary, high level measures that may cross multiple SCOR processes.

**Table 1.** SCOR Performance Attributes and Level 1 Metrics

| Performance Attributes | | | | | |
|---|---|---|---|---|---|
| Level 1 Metrics | Customer-Facing | | | Internal-Facing | |
| | Reliability | Responsiveness | Flexibility | Costs | Assets |
| Perfect Order Fulfillment | X | | | | |
| Order Fulfillment Cycle Time | | X | | | |
| Upside Supply Chain Flexibility | | | X | | |
| Upside Supply Chain Adaptability | | | X | | |
| Downside Supply Chain Adaptability | | | X | | |
| Supply Chain Management Cost | | | | X | |
| Cost of Goods Sold | | | | X | |
| Cash-To-Cash Cycle Time | | | | | X |
| Return on Supply Chain Fixed Assets | | | | | X |

Performance metrics equate to the BI key performance indicators (KPI). Each Level 1 metric can be decomposed to the lower Level 2 and Level 3 metrics, thus providing standardized performance measurement at the operational, tactical and strategic level across the supply chain.

## 2    Business Intelligence

For decades, corporate executives have made strategic business decisions based on information deduced from multiple reports that IT compiled by summarizing sets of frequently conflicting data.

Business intelligence systems promise to change that by, among other things, pulling data from all internal systems plus external sources to present a single version of the truth. This truth can then be delivered to decision-makers in the form of answers to highly strategic questions

Gartner, an information technology research firm, coined the term "business intelligence" during the 1990s. Business intelligence generally refers to the process of transforming the raw data companies collect from their various operations into usable information [7]. Since data in its raw form is of fairly limited use, companies are increasingly electing to use business intelligence software to realize their data's full potential. BI software comprises specialized computer software that allows an enterprise easily to aggregate, manipulate and display data as actionable information, or information that can be acted upon in making informed decisions.

By providing an insight into vital information, BI enables companies to improve the way they do business. Companies are empowered with the ability to offer products and services at the lowest possible cost and with the greatest amount of efficiency and productivity possible – while returning the highest revenues and profits.

Some companies are finding that it is beneficial to share BI capabilities with business partners as well as with employees. To do that, they are building Web-based "BI networks" to deliver intelligence to suppliers, consultants and others.

Business intelligence was once the domain of statisticians and corporate analysts. Not anymore. BI capabilities are spreading to virtually all parts of the organization, as companies strive to put critical data into the hands of business users who need it to do their jobs. Users want the following from their business intelligence systems [8]:

- The ability to run ad hoc queries

- Access to multiple databases

- Scalability, affordability and reliability

- Ease of integration with back-office systems

Many surveys conducted by the leading market research firms show that BI becomes one of the CIO's top priorities. While other segments of the enterprise software sector are floundering, interest and adoption in business intelligence continues to rise. CIOs recently surveyed by the IT research firm Gartner identified BI as their number-two technology priority for the coming year, a significant jump from the number-10 spot in 2004. The market is also on the rise. Forrester Research predicted that the BI reporting and analysis tools and applications software segment could scale up to $7.3bn by 2008, from $5.5bn in 2005 [9]. BI and business performance management account for a full 30 percent of the technology profile of a successful solution [10]. In the next five years we will see a dramatic 40% increase in the number of end users who use business-intelligence tools, and at least 50% of the Fortune 500 will turn to outsourcing contractors that have the next-generation technology and necessary expertise [11].

## 2.1    BI Challenges

Organizations are now willing to invest heavily in data warehousing software, servers and other hardware because they expect a rich pay-off. They anticipate that data warehousing will make huge numbers of employees more productive and efficient and result in better business decisions. However, doing this in practice is very difficult and is confronted with many challenges such as:

- Organizational and cultural differences
  Cultural and organizational issues can be attributed to the fact that supply chain processes are distributed among many internal and external organizational groups that tend to operate individually.

- Metrics
  Existing metrics do not capture how the overall supply chain has performed because they are primarily internally focused financial metrics [12]. The supply chain paradigm requires new metrics. The central place in the metrics system is taken by the Key Performance Indicators (KPI).

- Data quality
  One of the surveys showed that 75% of the organizations experienced financial pain from defective data [8]. Poor data quality costs them money in terms of lost productivity and faulty business decisions.

- Data security
  Security is one of the main IT concerns, since the information BI provides is the organization's most valuable asset. Fine-grained authorization and authentication, along with encryption are the requirements.

- Plenitude of data sources
  Most of the organizations have huge volumes of structured data housed in different data sources such as mainframes and databases, and also unstructured data sets. Providing the supply chain, integrating data from such a variety of sources is prerequisite for effective BI.

- Lack of expertise
  Experts knowledgeable in both SCM and data warehousing/BI are rare. Also, training is required for the business analyst and information workers in order to yield most benefits from SCI systems.

- End-user access
  The key to having a successful SCI system is having an interface that is simple to operate and offers personalization, customization, ad-hoc queries and collaboration through web portals.

### 2.1.1  BI Project Management

More than half of all Business Intelligence projects are either never completed or fail to deliver the features and benefits that are optimistically agreed on at their outset. While there are many reasons for this high failure rate. The biggest is that companies treat BI projects as just another IT project. It is, rather, a constantly evolving strategy, vision and architecture that continuously seeks to align an organization's operations and direction with its strategic business goals.

Organizations must understand and address these following challenges for BI success. BI projects fail because of [13]:

1. Failure to recognize BI projects as cross-organizational business initiatives, and to understand that as such they differ from typical standalone solutions.
2. Unengaged business sponsors (or sponsors who enjoy little or no authority in the enterprise).
3. Unavailable or unwilling business representatives.
4. Lack of skilled and available staff, or sub-optimal staff utilization.
5. No software release concept (no iterative development method).
6. No work breakdown structure (no methodology).
7. No business analysis or standardization activities.
8. No appreciation of the impact of dirty data on business profitability.
9. No understanding of the necessity for and the use of meta-data.
10. Too much reliance on disparate methods and tools.

### 2.1.2  BI Benefits

Implementing a data warehouse and using business intelligence and data mining technology can provide a significant benefit. For example, use it to:

- Analyze the performance and quality of the resource, for example, by comparing the process activity duration times across different resources.

- Understand and predict exceptions. BI can be used to understand the real cause of problems and, hopefully, avoid them based on knowledge gained from past process behaviour.

- Optimize processes. With BI, you can discover conditions under which specific paths or sub-paths of the process are executed, so you can redefine the process.

- Improve process execution times. Analyze process execution times and quality testing configurations of the system, assignment of resources and dynamic adaptation of the process.

## 2.2    BI Architecture

As the organization's information infrastructure expands across the Internet to encompass the entire supply chain, so does BI. The Internet expands the information sources of the data warehouse. It reaches beyond what is contained within the organization's internal systems, across the Internet, to include partner, supplier and customer systems.

In order to collaborate effectively, organizations must coordinate their businesses. According to the latest surveys, they are willing to share data and invest in collaboration. Some organizations have established a BI Competency Centre [14]. Team members should be cross-functional, full-time individuals who have the mission to identify and work with all BI-related initiatives within the enterprise. Building the BI system must begin with an understanding of the combined information needs of all users who might access the system.

The basic elements of a BI solution are: operational data store (ODS), data warehouse (with data and metadata), data mart (data warehouse that focuses on an individual subject area within the organization), ETL tools, OLAP engine, analytical tools (reporting, data mining, etc) and web portals. Combination of these elements forms a variety of possible scenarios which depend on the concrete organizational and informational structure of an organization and a supply chain.

Central to the BI system is a data warehouse. Generally, it can be centralized or distributed. It is important to make a distinction between the physical and the logical views. For example, a data warehouse can be centralized logically, but distributed physically thanks to distributed DBMS technology. The politics, the economics and the technology greatly favour a single centralized data warehouse [15].

A data warehouse system must:

- Make an organization's information easily accessible

- Present the organization's information consistently

- Be adaptive and resilient to change

- Be a secure bastion that protects information assets

- Serve as the foundation for improved decision making

- Have enterprise-wide support.

A data warehouse can be built using the top-down or the bottom-up approach [16]. Many organizations have built multi-tier warehouses (data warehouse with several data marts) using the bottom-up approach because of the lack of resources and the pressure to deliver solutions quickly. The top-down approach is better because it improves the consistency of information and reduces the number of extracts from operational systems. An organization can also choose the OLAP storage mode (Relational, Multidimensional, Hybrid) depending on the specific needs and priorities (real-time access, querying performance, storage, etc).

BI web portals are the ideal front-end because they utilize Internet technologies, offer personalization and customization, user-friendly, analytical tools and security mechanisms. The architecture of the supply chain BI system is shown in Figure 5 [17].

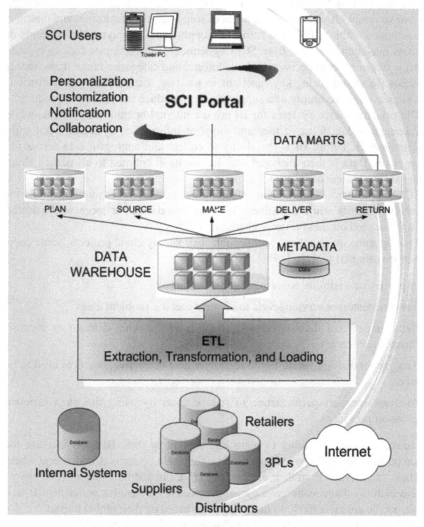

**Fig. 5.** Supply Chain BI Architecture

## 3    Supply Chain Intelligence

Supply Chain Intelligence (SCI) is a new initiative that provides the capability to reveal opportunities to cut costs, stimulate revenue and increase customer satisfaction by utilizing collaborative decision making [18]. SCI takes a broader, multidimensional view of the supply chain in which, using patterns and rules, meaningful information about the data can be discovered.

SCI technologies promise to extract and generate meaningful information for decision makers from the enormous amounts of data generated and captured by SCM systems.

The focus of SCM technologies primarily has been on providing operational and transactional efficiencies in the areas of sourcing, manufacturing, and distribution activities within a firm and across its supply chain. Applying the concepts of business intelligence to data from SCM systems, SCI technologies seek to provide strategic information to decision makers. Information categories range from what-if scenarios for reconfiguring key functions in sourcing, manufacturing and distribution to measuring the ability of a supply chain to produce cost-effective products.

The primary source systems for BI are the internal operational systems, while SCI integrates data from partner and supplier information systems. What truly differentiates SCI from BI is the ability to collect and aggregate data across the value chain. Data is then analyzed and the results distributed to all parties along that chain, regardless of location.

SCI complements supply chain planning because BI applications provide incremental benefits while a business lays the foundation for more sophisticated tools and related business process changes.

To reap some quick returns and support their supply chain projects, some companies are using BI tools to [19]:

- Improve data visibility so as to reduce inventory.

- Analyze customer service levels to identify specific problem areas.

- Better understand the sources of variability in customer demand to improve forecast accuracy.

- Analyze production variability to identify where corrective measures need to be taken.

- Analyze transport performance to reduce costs by using the most efficient transport providers.

By providing wider visibility to plans and supporting data, BI tools increase the return on existing SCP applications because they help companies understand where and how they deviate from their planned objectives. In addition, they provide shared data availability that encourages a global perspective on business performance. As a result, people are more likely to make decisions based on their global impact.

Organizations can apply BI in the following supply chain areas [20]:

- Plan Analytics — balancing supply chain resources with requirements.

- Source Analytics — improving inbound supply chain consolidation and optimization.

- Make Analytics — providing insight into the manufacturing process.

- Deliver Analytics — improving outbound supply efficiency and effectiveness.

- Return Analytics — managing the return of goods effectively and efficiently.

For example, demand forecasting is one of the key applications of data mining. Complex demand forecasting models can be created using a number of factors like sales figures, basic economic indicators, environmental conditions, etc. If correctly implemented, a data warehouse can significantly help in improving the retailer's relations with suppliers and can complement the existing SCM application.

SCI can have the following industrial applications in the area of supply chain management:

- Sales/Marketing. Providing analyses of customer-specific transaction data. Enabling retailers to know not only what's selling but who is buying it. Strengthening consumer 'pull'.

- Forecasting. Using scanning data to forecast demand and, based on the forecast, to define inventory requirements more accurately.

- Ordering and replenishment. Using information to make faster, more informed decisions about which items to order and optimum quantities.

- Purchasing/Vendor Analysis. Helping purchasing managers to understand the different cost and timeliness factors of each of their suppliers.

- Distribution and logistics. Helping distribution centres manage increased volumes. Can use advance shipment information to schedule and consolidate inbound and outbound freight.

- Transportation management. Developing optimal load consolidation plans and routing schedules.

- Inventory planning. Helping identify the inventory level needed, item by item, to ensure a given grade of service.

- Stock location planning. Helping warehouse planners assign products to locations so as to minimize distances, improve efficiency.

- Finished goods deployment. Balancing supply, demand, and capacity to determine how to allocate limited quantities of finished goods.

## 3.1    BI and Logistics

Increased impetus on core competence, globalization, and the emergence of the Internet has given rise to a new breed of e-Logistics companies called Third Party

Logistics-3PL providers, which offer a spectrum of solutions [21]. They need to establish themselves as key business partners involved in the entire supply chain – right from logistics strategy formulation to its implementation. And they need to effectively share information and knowledge with the customers using the BI tools. Business Intelligence can help the 3PLs in the following ways [22]:

- Service Improvement: Business Intelligence can improve the effectiveness of the logistics services by in-depth analysis and reports on various functions involved in these services.

- Provide Information Technology Based Services: With the help of BI, 3PLs can provide their clients with analysis and reports specific to their supply chain. These can significantly help the customers increase their responsiveness and time to market.

- Improve Organizational Support Functions: BI can significantly improve organizational support functions like HR and financial management by providing an integrated view of these functions and supporting their specific decision making requirements.

## 4    Service-Oriented Business Intelligence

As companies move from simply monitoring to proactively managing business performance, they need real-time visibility in market, customer and competitive conditions. This requires integration across IT systems and business processes in a cost-effective and flexible manner.

A service-oriented architecture (SOA) provides a standardized, flexible approach to enterprise business process integration and role-based access to process activities, information (including BI), business transactions and collaboration tools. A SOA makes it possible to separate processes from applications and to create on-demand and event driven information, application and collaborative services that can be invoked in an industry standard way. These services can then be rapidly assembled into composite applications associated with individual process activities in an industry standard way.

There are many reasons why companies are investing in SOA initiatives at present. The obvious one is that a SOA reduces operational costs and increases efficiency, improves effectiveness and increases collaboration. Improving efficiency is typically achieved by standardizing on common business processes that have been separated from applications and then integrating and automating them by mapping process activities to applications within the enterprise and across businesses. A SOA can be used to improve effectiveness by leveraging business intelligence everywhere in order to guide employees during every performed activity so that everyone contributes to strategic objectives and executes on a common business strategy. It is also possible to monitor events to automatically detect/ predict problems and opportunities for rapid response and business optimization.

Improving collaboration allows employees, partners, customers and suppliers access to team workspaces where they can share information and services.

## 4.1    Web Services and BI

The kind of BI, performance management and data integration artefacts that can be developed and published as web services include [23]:

- Queries
- Reports
- OLAP slice services (MDX queries)
- Scoring and predictive models
- Alerts
- Scorecards
- Plans
- BAM agents
- Decisions (i.e., automated decision services)
- Data integration workflows and federated queries

Figure 6 shows the idea of composite SOA with BI services being accessible for portals, processes, operational applications, performance management scorecards, search engines and office applications all via an Enterprise Service Bus (ESB).

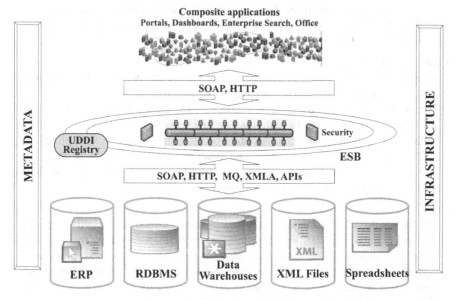

**Fig. 6.** BI SOA

A key point to note is that a SOA can interconnect resources that operate at the user interaction, application and data levels of an IT system. In a data warehousing environment, for example, a data validation service could be defined as a service provider and be called by a data integration application. Other examples of services providers include user authentication, search, data transformation, BI analysis, data mining models, legacy applications and business transactions.

The advantage of a SOA is that it enables common and shared interfaces to be defined and created for distributed resources. This eliminates the need to develop multiple proprietary point-to-point connections between resources, which reduces IT development and maintenance efforts and encourages service reuse [24].

One of the potential users of an ESB is a data integration service. In fact, several vendors have modified their data integration and ETL (extract, transform and load) tools to be event-driven so that they can consume event messages from an ESB. These events can carry information about source data changes, which can be used by the data integration service to incrementally update an operational data store (ODS) or data warehouse.

The Impact of Web Services on business intelligence systems can be seen at the following three levels [25]:

- Data Level
  Web services will affect the access and delivery of BI data in several ways:
  - First, the input and output interfaces will evolve from proprietary API mechanisms to standard-based mechanisms, such as XML for Analysis, XBRL and JOLAP, that ease BI data integration.
  - Second, authentication tools, such as LDAP, NTLM and ADS, will help ensure authorization rights of users to data across applications.
  - Finally, encryption from Secured Socket Layers (SSLs) and Hypertext Transfer Protocol Secure (HTTPS) will help protect the integrity of a message during transit.

- Metadata Level
  Web services will also have a significant impact on BI metadata. The functionality of defining dimensions, hierarchies, calculations, business rules and reporting formats and sharing them across BI tools and applications will be exposed as a Web service. In addition, the functionality of modifying metadata definitions, versioning the changes and synchronizing them across applications will be exposed as a Web service. And again, the import and export interfaces will evolve from proprietary mechanisms to XML-based mechanisms such as CWM in order to ease BI metadata integration.

- Process Level
  Web services will be a major technology enabler for business activity monitoring and collaborative BI. By encapsulating application functionality and business rules within a Web service, companies will be able to create intelligent agents that monitor events in real-time, dynamically route information to targeted users or processes and automate analysis to improve the speed and effectiveness of adapting business operations to changing conditions.

## 4.2    Service-Oriented Database Architecture

Service-Oriented Database Architecture (SODA) closely mirrors modern practice in application construction and allows for unlimited scalability by dividing database processing along service boundaries. Services can be scaled independently or partitioned into new services to handle additional load, availability, or business requirements. Each Service can be made highly available, and the overall application can be designed to provide continuous availability [26].

Unlike transparent scale-out, SODA avoids SQL-level cross-partition operations, and all the scalability limitations they bring, in favour of well-defined requests between database services. Additionally, unlike non-transparent scale-out, SODA integrates support for service interfaces and inter-service communications, and also routing directly into the database system, thus relieving the application development and maintenance burden.

The notion of a database service is central to SODA and its scalability model. At a logical level, a database service exposes a well-documented application level interface to data. This is not a general database interface for reading and writing data, but instead provides very specific application functionality. For example, an inventory database service might expose methods for checking inventory levels, reserving inventory, removing products from inventory, recording receipt of new shipments and managing back-ordered items.

The first difference between database services and traditional models is that access to data under the control of one database service is completely isolated from access to data under the control of a different database service.

The second difference is that requests to database services are not made over a database connection but, instead, the services are exposed as Web Services.

Database services encourage far more business logic to be managed by the database system than has classically been done. SODA supports the use of general purpose programming languages for creating business rules inside the database system.

With the SODA, databases are partitioned according to well-defined service boundaries that meet application requirements. These database services can then be hosted on a single server node, or multiple server nodes, to achieve the desired level of scalability.

# 5    Business Activity Monitoring

Most businesses are probably not using BI to continually and automatically monitor events in their operational business processes as their businesses operate to respond rapidly to detected problems or to predict whether problems lie ahead. In general, therefore, companies have no active real-time element to their BI systems. The consequences are that nothing is helping the business to automatically respond immediately when problems occur or opportunities arise. Also, there is no automatic notification or flagging of alerts to take action that may avoid unnecessary costs, business disruption, operational mistakes and unhappy customers in the future.

Business Activity Monitoring (BAM) is a collection of tools that allow you to manage aggregations, alerts, and profiles to monitor relevant business metrics (Key Performance Indicators - KPIs). It gives users end-to-end visibility into business processes, providing accurate information about the status and results of various operations, processes, and transactions so they can address problem areas and resolve issues within your business. BAM software products incorporate concepts from — and sometimes are built on — ERP, business intelligence, business process management and enterprise application integration (EAI) software.

The BAM provides an easy, real-time, transaction-consistent way to monitor heterogeneous business applications and to present data for SQL queries and aggregated reports (OLAP). Through queries and aggregations BAM systems can include not only the data that is present during the running business process, but also the state and the dynamics of the running business process, independent of how the business is automated.

Figure 7 shows how data and messages flow within the BAM system.

BAM applies operational business intelligence and application integration technologies to automated processes to continually refine them based on feedback that comes directly from knowledge of operational events [27]. In addition to auditing business processes (and business process management systems), BAM can send event-driven alerts that can be used to alert decision makers to changes in the business that may require action.

Possible business reasons for the BAM system deployment are [28]:

- To detect any changes in orders (cancelled orders, large orders, orders from valuable customers) and, if need be, to alert the sales force and other people in operations when specific events have a business impact, e.g., to respond when cancelled orders occur.

- To check orders or predicted demand against inventory to optimize a supply chain by reordering/delaying/cancelling supply of inventory.

- To detect or predict bottlenecks in the package assembly activity.

- To detect or predict delays in shipments for valuable customers.

- To detect or predict cash flow problems because of late payments.

Business users use Business Activity Monitoring to gain a real-time holistic view of business processes that span heterogeneous applications, regardless of the infrastructure implementation.

BAM gives a different perspective on a business process. For example, a BAM system might provide graphical depictions of per-product sales trends or current inventory levels or other key performance indicators. The information might be updated every day, every hour, or more frequently.

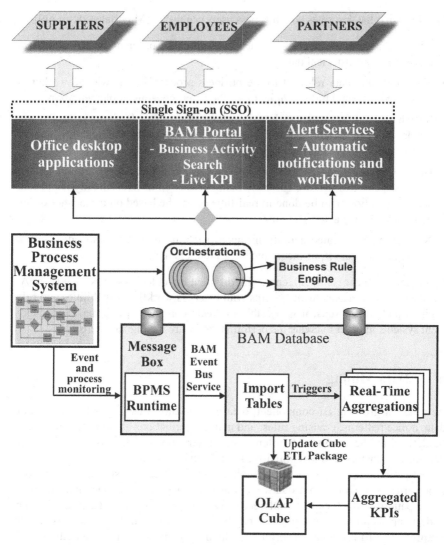

**Fig. 7.** BAM Architecture

BAM relies on one or more BAM activities. A BAM activity represents a specific business process, such as handling purchase orders or shipping a product, and each one has a defined set of milestones and business data. For example, a purchase order activity might have milestones such as Approved, Denied, and Delivered along with business data like Customer Name and Product.

One of the newest technology trends is toward the integration of BAM systems with web portals. Business end users can use the BAM portals to monitor KPIs, which measure progress toward a business goal, as well as other information about their business process.

The following list describes how business users can use BAM portal systems [29]:

- View a single activity instance such as a purchase order or loan (process) in real-time or as historical data.

- Show only the data relevant to the business process the knowledge worker is concerned with and hide complexity of the heterogeneous implementation.

- Search for activity instances based on their progress or business data. For example, you can search for loans that are waiting for customer signature and the dollar amount is greater than a given value.

- Browse aggregations (which are key performance indicators) around all the business activities that are currently being processed or have already happened. The aggregations can be done in real-time or can be based on a snapshot of the activities taken at a specific time.

- Navigate to the related activity instances such as shipments associated with a given purchase order, or the invoice in which it is included.

By combining business intelligence with different novel technologies such as SOA, business process management systems with business workflow capabilities and rule engines, and web portals, it is possible to create composite performance management systems in order to create the performance-aware intelligent enterprise.

# 6    Data Mining

As a fastest growing BI component, data mining allows us to comb through our data, notice patterns, devising rules, and making predictions about the future It can be defined as the analysis of (often large) observational data sets to find unsuspected relationships and to summarize the data in novel ways [30].

Data mining applies algorithms, such as decision trees, clustering, association, time series, and so on, to a dataset and analyzes its contents. This analysis produces patterns, which can be explored for valuable information. Depending on the underlying algorithm, these patterns can be in the form of trees, rules, clusters, or simply a set of mathematical formulas. The information found in the patterns can be used for reporting, as a guide to supply chain strategies, and, most importantly, for prediction [31].

Data mining can be applied to the following tasks [32]:

- Classification
- Estimation
- Segmentation
- Association
- Forecasting
- Text analysis

The main characteristics of data mining intelligent applications are:

- Make decisions without coding – Data mining algorithms learn business rules directly from the data, freeing you from trying to discover and code them yourselves.

- Customized for each client – Data mining learns the rules from the client's data resulting in logic that is automatically specialized for each individual client.

- Automatically update themselves – As client's business changes, so do the factors that impact their business. Data mining allows application logic to be automatically updated through simple processing steps. Applications do not need to be rewritten, recompiled or redeployed and are always online.

Data mining provides answers to three key business questions as shown in Figure 8.

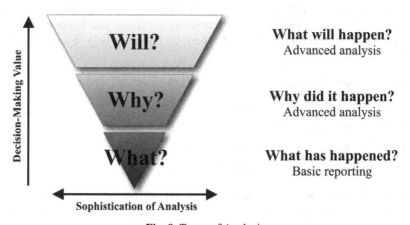

**Fig. 8.** Types of Analysis

Data mining techniques can be applied to many applications, answering various types of businesses questions, such as demand forecasting, inventory prediction, customer and product segmentation, risk management, etc.

Most of the data mining techniques existed, at least as academic algorithms, for years or decades. However, it is only in the last decade that commercial data mining has caught on in a big way. This is due to the convergence of several factors [33]:

- The data is being produced.
- The data is being warehoused.
- Computing power is affordable.
- Interest in customer relationship management is strong.
- Commercial data mining software products are readily available.

The business intelligence market has shifted its focus from the IT community to the broader line of business users, we can expect data mining to do the same.

Over the next few years, data mining technology should significantly grow, thanks to business applications and database software vendors.

Embedding data mining algorithms directly into business applications is a perfect conduit to reach them. As vendors create a closed loop between their data warehouses and their enterprise business processes, this will extend to areas like finance and the supply chain.

Market research groups expect the data mining market to expand 10% annually over the next few years [34].

# 7    Business Intelligence Web Portals

Technologies such as portals comprise a portfolio of collaboration and communication services that connect people, information, processes and systems both inside and outside the corporate firewall. These IT resources help to solve key information worker challenges, such as:

- Finding information more easily
- Working more productively in teams
- Connecting more effectively with others
- Collaborative planning and decision making
- Providing a consistent user experience.

Industry analysts predict that collaboration is rapidly becoming a strategic platform advantage, yet many organizations either already have or will implement such capabilities at the tactical level, purchasing and deploying collaboration technologies in a piecemeal fashion. The end result is likely to be expensive systems that are poorly integrated and costly to maintain.

Organizations usually view portals and business intelligence separately, but in the next few years, the two will become integrated. These technologies can shrink the amount of data that has to be analyzed to make a decision, in an era when people have less time and must focus on the decision criteria that truly make a difference [8].

The true value of the combination of BI tools and portal technology is that decision makers will have more complete information integrated on their screens from across their enterprise and their partners to make better decisions more quickly.

Portals provide the ideal framework for the integration of critical business intelligence information because portals support the delivery of a highly secure, unified access point through consistent interfaces to Web applications spanning a range of applications and systems [35].

Additionally, portals provide valuable functions like search, collaboration and workflow in a security-rich environment - as well as new capabilities that enable different kinds of analytics.

A BI web portal can be viewed as an integrated, web-based online analytical processing solution that enables employees throughout the entire supply chain to create and share reports, charts and pivot tables, perform ad-hoc analysis, based on online OLAP services, cube files, relational databases and web services.

Supply chain BI web portals combine modern technologies like portlets (web parts), content management, enterprise search and web services in order to provide better user experience for collaborative decision making.

In this way, it is possible to present different KPIs combined in a form of digital dashboard in order to provide a comprehensive view to the decision makers.

The BI web portal typically consists of many modules (web parts) with different functions. There can be web parts for organizing content structure, for creating views of data, for data analysis (querying, drill-down, cell colouring), for document exchange, etc.

Figure 9 shows a BI portal with two data views that introduce two reports for presenting data mining results. The reports are stored in a separate report server and integrated in the portal using standard XML web service and SOAP (Simple Object Access Protocol) technologies.

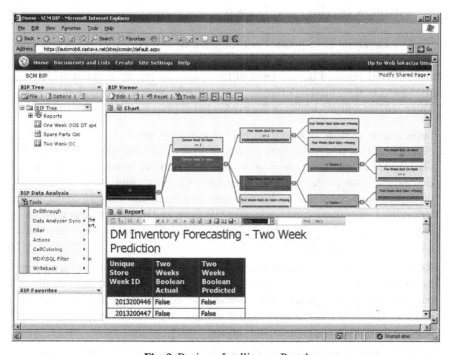

**Fig. 9.** Business Intelligence Portal

Another usage of web portals is related to the Balanced Scorecard (BSC) methodology. Web portals can bring visibility to the BSC process, ease a cultural transition, and enable participation by a wider audience. Many BI software vendors are adding BSC capabilities to their product line.

The main characteristics of the BI web portal are:

- Modularity – ability to compose web pages dynamically integrating content from different sources and using different technologies.

- Personalization – portal can be easily personalized to reflect users' needs and interests.

- Customization – Users can tailor a web portal to suit their particular needs.

- Self-service – Users can create new modules, site templates, share and reuse proven solutions, and get or access information on-demand.

- Fine-grained security mechanisms – role-based access control, authentication and authorization mechanisms and granular security.

- Easy user adoption – standardized and consistent browser-based interface enables users to access information using the familiar tools.


# 8    Business Intelligence Trends

As financial analysts have pointed out [36], the BI software segment is a bright spot in the software industry and continues to grow even in a tough economic climate. The BI segment is composed of firms specializing in BI and those offering BI as part of their product offerings such as ERP and database vendors.

There are three strong trends in the market today [37]:

- Solid market growth – Overall BI market growth is solid. At this point, however, the market is splitting into a group of haves and have-nots. Many smaller vendors are experiencing slowing growth rates and are bleeding money. This creates a vicious circle. Potential customers shy away from them because of their poor financial position, limited customer base, and concern about their longevity. Under these conditions, many BI firms and their customers are looking for an exit strategy with a larger firm acquiring them and incorporating their products into their offerings.

- Ripe for consolidation – A mature market like BI with many competitors is ripe for consolidation. Stronger players acquire smaller players for their customer base, technology and people. It is assumed that BI software companies need to reach $1 billion in sales to have sufficient depth to continue expanding and enhancing their products, support and services.

- Pressure from all sides – BI vendors are under pressure from ERP vendors at the high end and from open-source BI initiatives at the low end.

High end – In a effort to jumpstart their growth and capture more client dollars, ERP vendors like SAP, Oracle and Microsoft have expanded their reporting offering to include data warehousing, business intelligence, and analytics applications. Their attempts to win clients threaten the expansion of the pure BI vendors. Although some ERP vendors have included pure BI vendors' tools in their analytic offerings, this still jeopardizes the BI vendors.

Low end – From the low end of the marketplace, open-source products are going to be key competitors. Open source is gaining momentum and there are significant amounts of money coming from the investment community into "open source" vendors. Open source approaches benefit both customers and vendors. For the customer, the benefit is lower software costs. The downside, however, can be higher development costs.

Business intelligence and data warehousing (DW) has reached a new level of maturity, as both a discipline and a technology market. Demand for BI/DW is stronger than ever and BI/DW are within the top ten CIO priorities. Most enterprises already have a BI/DW infrastructure in place and are now taking the lessons they have learned from previous efforts to remedy problem areas. At the same time, many enterprises are also moving towards the next steps in the evolution of BI/DW.

Besides the already discussed cutting-edge BI technologies and trends such as service-oriented architecture, business activity monitoring, predictive analytics and web portals, there are also some other important business intelligence trends that are shaping BI/DW today, as well as the new technologies and initiatives that are moving BI/DW forward.

- Infrastructure standardization and consolidation
  Enterprises tend to know what they are spending for ERP and other core systems, but not for BI/DW. That's because BI/DW efforts have largely been undertaken in silos, with each business domain creating its own solutions to the problem of obtaining and analyzing data. This siloed approach almost always results in duplication of effort, inefficiency and increased expense.

  Enterprises have come to recognize their disparate BI/DW solutions as a problem over the past couple of years. Their interest has been particularly piqued in these lean economic times, when eliminating duplicate BI tools or data marts might result in lower license costs and maintenance expense. Improved access to information, while more difficult to quantify, is also an important benefit of eliminating silos. However, standardizing and consolidating a BI/DW infrastructure is far easier said than done. It involves political and organizational issues that are just as challenging as the technology issues.

- Metadata and master data management
  Within every enterprise, there is a set of data that provides valuable information to identify and uniquely define core entities, such as customers, products, suppliers, etc.

  The proliferation of enterprise applications, combined with most organizations' siloed approach to BI/DW, has resulted in master data being scattered across the enterprise. The drive toward integrating and streamlining enterprise systems is getting more and more attention.

Another increasingly important feature is metadata management, which enables information consistency. The Common Warehouse Metamodel (CWM) is an open industry standard defining a common metamodel and XML-based interchange format for meta data in the data warehousing and business analysis domains [38].

- The rise of the BI application service providers
  Given the cost and difficulty of developing and implementing BI solutions, by 2010, at least 50% of the Fortune 500 will turn to outsourcing contractors that have the next-generation technology and database marketing expertise to do it [39].

  The worldwide software-as-a-service (SaaS) market reached $6.3 billion in 2006 and is forecast to grow to $19.3 billion by year-end 2011, according to Gartner [40]. SaaS is hosted software based on a single set of common code and data definitions that are consumed in a one-to-many model by all contracted customers, at any time, on a pay-for-use basis, or as a subscription based on usage metrics.

- Web 2.0 and BI 2.0
  Essentially, Web 2.0 is an umbrella term for a group of technologies that have advanced web usage and turned the web into a development platform for the enterprise. These technologies include: RSS and ATOM feeds, web services, JavaScript and AJAX, web scripting, mashups, programming frameworks (e.g., Adobe Flex, Ruby on Rails, and OpenLaszlo) and Wikis.

  Currently, we are about to enter a new era of Rich Internet Applications (RIA), when plain looking Web applications will gradually be replaced with RIA delivered over the Web [41]. RIA applications run in a virtual machine deliver over the Web and have a potential of becoming full featured desktop applications. This new approach will be part of a "Business Intelligence 2.0" revolution that makes any company's data more democratic [42].

- Adaptive Business Intelligence
  There is also a new trend emerging in the marketplace called Adaptive Business Intelligence. Adaptive Business Intelligence can be defined as the discipline of combining prediction, optimization and adaptability into a system capable of answering these two fundamental questions [43]: What is likely to happen in the future? and What is the best decision right now?

  This relatively new approach to business intelligence is capable of recommending the best course of action (based on past data), but it does so in a very special way: an Adaptive Business Intelligence system incorporates prediction and optimization modules to recommend near-optimal decisions and an "adaptability module" for improving future recommendations. Such systems can help business managers make decisions that increase efficiency, productivity, and competitiveness.

- Simulation and Business Intelligence

Modelling and analysis of supply chains can be extremely challenging due to a complex network structure, process relationships, constraints and especially uncertainty. Simulation can be a valuable tool for studying the behaviour of a supply network model under the most realistic business conditions.

The new approach to supply chain simulation allows process-based modelling of any supply chain configuration regardless of the number of supply chain nodes, process levels, process types, constraints, or business policies involved [44]. Also, the database-centric approach for supply chain simulation enables all the relevant data to be stored (models, processes, relationships, constrains, metrics, etc.).

The huge amount of data generated through simulation runs, now kept in the database, can be loaded into the data warehouse for further analysis and data mining. This way, different scenarios and business policies can be evaluated, KPIs can be monitored and predictions about the future can be made.

# 9    Conclusion

Market forces are driving the need for collaboration. Customers are expecting more and the economy is demanding greater cost efficiencies. Organizations working together as part of a collaborative supply chain have understood the need for better information exchange.

To succeed in a competitive marketplace, an agile supply chain requires business intelligence (BI) systems to quickly anticipate, adapt, and react to changing business conditions. BI systems provide sustainable success in a dynamic environment by empowering business users at all levels of the supply chain and enabling them to use actionable, real-time information.

Supply chain intelligence (SCI) reveals opportunities to reduce costs and stimulate revenue growth and it enables companies to understand the entire supply chain from the customer's perspective.

Many forward-thinking supply networks have realized that business intelligence networks are a first step to information consolidation and gaining visibility over the value chain. With BI networks, businesses can share information with customers, suppliers and partners.

Thanks to a new emerging set of Internet-based technologies such as web services and SOA, businesses are taking BI networks to the next level. These next-generation and loosely-coupled networks, built as BI web services, will enable collaborative, efficient, responsive and adaptive supply chains.

Furthermore, by combining business activity monitoring (BAM) in support of process management and event monitoring together with web portals for flexible, user-friendly and accessible information delivery, on top of the BI systems, it is possible to create a synergistic effect.

The way organizations gather, measure and analyze information often determines the ultimate supply chain success.

In this chapter, we discuss the latest supply chain management issues and the drivers for the implementation of business intelligence systems and performance measurement based on the process approach.

We also talk about BI challenges and benefits, as well as applications of BI technologies in the supply chain management domain.

Finally, the main BI trends and advanced IT technologies that will shape future BI systems are introduced.

# References

1. Martin, J., Roth, R.: Supply Chain management – Direction Strategy, ECRU Technologies, Inc. (2000)
2. Lambert, M.D., Cooper, C.M., Pagh, D.J.: Supply Chain Management: Implementation Issues and Research Opportunities. The International Journal of Logistics Management 44(2), 1–19 (1998)
3. Gintic, Measuring supply chain performance using a SCOR-based approach, Institute of Manufacturing Technology (March 2002)
4. Lapide, L.: What About Measuring Supply Chain Performance? ASCET 2 (2000)
5. Supply Chain Council, Operations Reference-Model Overview Version 8.0 (2006)
6. Bolstroff, P., Rosenbaum, R.: Supply Chain Excellence: A Handbook for Dramatic Improvement Using the SCOR Model. Amacom, New York (2003)
7. Quinn, K.: Establishing a Culture of Measurement – A Practical Guide to Business Intelligence, Information Builders (2003)
8. Computerworld, Executive Briefings - Get Smart About Business Intelligence (2005)
9. Datamonitor, BI Trends – What to Expect in 2006 (January 2006)
10. Decker, J., Brett, C.: The Joy of SOX: Part 2—The SOX Solution Blueprint, META Group (2003)
11. Betts, M.: The future of business intelligence, Computerworld (2003)
12. Lambert, M.D., Pohlen, L.T.: Supply Chain Metrics. The International Journal of Logistics Management 12(1), 1–19 (2001)
13. Atre, S.: The Top 10 Critical Challenges for Business Intelligence Success, Computerworld, Computerworld (2003)
14. Biere, M.: Business Intelligence for the Enterprise. Pearson Education, New Jersey (2003)
15. Inmon, H.W.: Building the Data Warehouse, 3rd edn. John Wiley & Sons, Chichester (2002)
16. Moeller, A.R.: Distributed Data Warehousing Using Web Technology. Amacom, New York (2001)
17. Stefanovic, N., Radenkovic, B., Stefanovic, D.: Supply Chain Intelligence. In: Pham, D.T., Eldukhri, E.E., Soroka, A.J. (eds.) Intelligent Production Machines and Systems, vol. 3, Whittles Publishing, Dunbeath (2007)
18. Haydock, P.M.: Supply Chain Intelligence. ASCET 5, 15–21 (2003)
19. Shobrys, D.: Supply Chain Management and Business Intelligence, Supply Chain Consultants (2003)

20. Curt, H.: Supply Chain Intelligence: Applying Business Intelligence to Enhance Operational Efficiencies, Wipro (2002)
21. Wolfe, M.E., Wadewitz, R.T., Combe, G.C.: E-gistics, Bear, Stearns & Co. Inc. (2000)
22. Srinivasa, P.R., Saurabh, S.: Business Intelligence and Logistics, Wipro (2001)
23. Ferguson, M.: Developing a Service-Oriented Architecture (SOA) for Business Intelligence, BeyeNetwork (2007)
24. White, C.: What Do SOA and ESB Mean in Business Intelligence, BeyeNetwork (2007)
25. Everett, D.: Web Services and Business Intelligence. Hyperion, New York (2003)
26. Berenson, H.: Why Consider a Service-Oriented Database Architecture for Scalability and Availability, Microsoft (2005)
27. Microsoft, What is BAM? (March 2006), `http://msdn2.microsoft.com/en-us/library/aa560139.aspx`
28. Ferguson, M.: Building Intelligent Agents Using Business Activity Monitoring. DMReview Magazine (Dec. 2005)
29. Stefanovic, N., Stefanovic, D.: Methodology for BPM in Supply Networks. In: 5th CIRP International Seminar on Intelligent Computation in Manufacturing Engineering, Ischia, Italy (2006)
30. Hand, D., Mannila, H., Smyth, P.: Principles of Data Mining. MIT Press, Cambridge (2001)
31. Tang, Z.H., MacLennan, J.: Data Mining With SQL Server 2005. Wiley, Indianopolis (2005)
32. Larose, T.D.: Discovering Knowledge in Data. John Wiley & Sons, Chichester (2005)
33. Berry, M.J.A., Linoff, G.S.: Data Mining Techniques for Marketing, Sales, and Customer relationship Management. John Wiley & Sons, Chichester (2004)
34. Beal, B.: Application Vendors to Dig Into Data Mining (Jan. 2005), `http://search-crm.techtarget.com/originalContent/0,289142,sid11_gci10473 47,00.html`
35. Bisconti, K.: Integrating BI Tools into the Enterprise Portal. DMReview Magazine (Aug. 2005)
36. Athena IT Solutions, BI as a Smart Investment (2006), `http://www.athena-solutions.com/bi-brief/june03-issue3.html`
37. McKnight, W., Humphrey, S.: Building Business Intelligence: Rafting Into the Business Intelligence Future. DMReview Magazine (Oct. 2004)
38. OMG, Common Warehouse Metamodel, CWM (2007), `http://www.omg.org/technology/documents/formal/cwm.htm`
39. Computerworld, The Future of Business Intelligence (June 2004)
40. Linthicum, D.: Gartner Sees $19.3 Billion SaaS Market by 2011 (August 2007), `http://www.intelligententerprise.com/blog/archives/2007/08/gartner_sees_19.html`
41. Flex News Desk, Business Intelligence in the world of Rich Internet Applications (2007), `http://java.sys-con.com/read/280900.htm`
42. Ames, B.: Web 2.0 tools inspire data-sharing software (2007), `http://www.infoworld.com/article/07/04/18/HNweb2datasharingtools_1.html`
43. Michalewicz, Z., Schmidt, M., Michalewicz, M., Chiriac, C.: Adaptive Business Intelligence. Springer, Heidelberg (2006)
44. Stefanovic, D., Stefanovic, N.: Methodology for modeling and analysis of supply networks. Journal of Intelligent Manufacturing 19(4), 485–503 (2008)

# Author Index